FLAK 88

8.8cm Flugzeugabwehrkanone (Models 18/36/37/41)

COVER IMAGE:
8.8cm Flugzeugabwehrkanone *(Mark Rolfe)*

© Chris McNab 2018

All rights reserved. No part of this publication may be reproduced or stored in a retrieval system or transmitted, in any form or by any means, electronic, mechanical, photocopying, recording or otherwise, without prior permission in writing from Haynes Publishing.

First published in March 2018

A catalogue record for this book is available from the British Library.

ISBN 978 1 78521 133 1

Library of Congress control no. 2017948766

Published by Haynes Publishing,
Sparkford, Yeovil, Somerset BA22 7JJ, UK.
Tel: 01963 440635
Int. tel: +44 1963 440635
Website: www.haynes.com

Haynes North America Inc.,
859 Lawrence Drive, Newbury Park,
California 91320, USA.

Printed in Malaysia.

Commissioning editor: Jonathan Falconer
Copy editor: Michelle Tilling
Proof reader: Penny Housden
Indexer: Peter Nicholson
Page design: James Robertson

Acknowledgements

I would like to give sincere thanks to those collectors who provided me with access to well-restored 8.8cm guns and related equipment for photographic purposes. Specifically, thanks to Bruce and Max Crompton of Axis Track Services (www.axistrackservices.com) and Sir Michael Savory of the Muckleburgh Military Collection (http://www.muckleburgh.co.uk). Photographs taken by the author from these collections are acknowledged in the text. I would also like to thank John Baum (www.germanmanuals.com) for his kind permission to use text from his translations of rare German 8.8cm manuals.

FLAK 88

8.8cm Flugzeugabwehrkanone (Models 18/36/37/41)

Owners' Workshop Manual

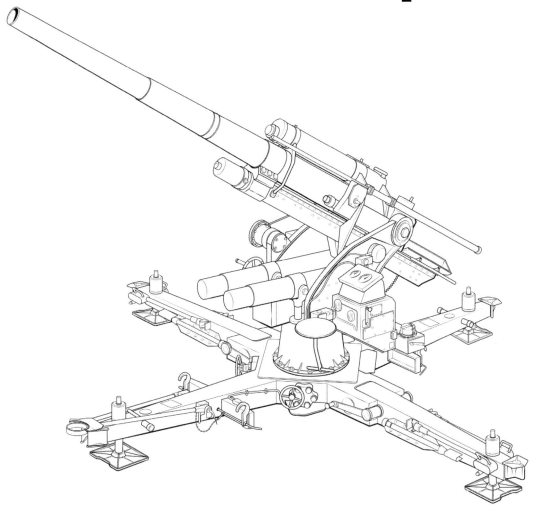

The most deadly and successful artillery gun in the
Wehrmacht arsenal in the Second World War

Chris McNab

Contents

| 6 | Introduction |

| 8 | Development |

Flak 18, Flak 36 and Flak 37 — 12
The Flak 41 — 18
Diversity — 21

| 28 | Design and basic operation |

The barrel and breech — 30
Recoil mechanism — 35
Mount — 38
Sighting and fire control — 47

| 56 | Ammunition |

Shell cases — 58
Propellants — 60
Fuzes — 62
Warheads — 64
Pak 43 and tank gun ammunition — 71

| 74 | Crew roles and gun operation |

Manning the 8.8cm Flak — 80
Handling the gun in action — 84
Theatre challenges — 90

| 100 | At war – anti-aircraft operations |

Organisation of Flak units — 103
Controlling the fire — 110
The effectiveness of the 8.8cm Flak — 119

OPPOSITE A Flak 36 team conducts an exercise in Germany in September 1937. This image gives a good impression of the configuration of a field anti-aircraft post, with the crew staying hunkered down behind substantial front cover. *(Alinari Archives/Getty Images)*

ABOVE A view of the right side of the 8.8cm Flak, showing traverse and elevation mechanisms and the Übertragungsgerät 37 receiver dials. *(Author/Muckleburgh Collection)*

| 122 | At war – anti-tank operations |

North Africa and the Western Desert — 126
The Eastern Front — 133
Fall in the West — 138

| 148 | Transportation and maintenance |

Half-track prime movers — 150
Inspection, troubleshooting and maintenance — 154
Diagnostics — 161

| 166 | Endnotes |

| 167 | Appendices |

Bibliography and further reading — 167
8.8cm Flak 36 gun specifications — 168

| 170 | Index |

Introduction

It is actually quite rare that an artillery piece gains a high level of notoriety or public awareness. Such a status is usually conferred on guns of prodigious size, from the monstrous 42cm 'Big Bertha' of the First World War through to Saddam Hussein's attempted 'Supergun'. Elevating a standard production weapon is a far rarer thing, and yet this has been achieved by the 8.8cm Flak guns (the Flak 18/36/37/41), probably to a greater degree than any other similar weapon system.

This status is not the post-war imposition of historians. Indeed, during the Second World War (1939–45) the '88' was feared and admired by the Allies and trusted and applied by the Germans in every theatre in the Western hemisphere. A first-hand account[1] clarifies the point – here the words of one Private First Class Lewis of the 82nd Airborne Division, remembering an attack by the US paratroopers on the German positions at Thier-du-Mont, France, in 1944, supported by five M10 tank-destroyers:

Suddenly one of our soldiers shouted, 'My God, look at that!' I was terrified as I observed three well-camouflaged barrels of the powerful German 88mm gun about two hundred [and] fifty feet in front of us. These guns were defended by infantry in foxholes, with several machine guns scattered along the ridge.

'Bazookas up front!' Captain Wilde ordered. At that same moment the German 88s opened fire. Blam! Blam! Just like that four of the M10s were hit and destroyed. One of the M10s caught fire and the ammunition it had stored exploded with a huge, fiery blast. Along with another bazooka man, Private Gerald D. Jones, I ran forward. We both could see a German 88 hidden in the woods located along a fence row. The gun crew was working feverishly to load and fire their gun as fast as they could.

We both took aim at the 88 gun as our loaders worked frantically to load our weapons. The Germans could see us preparing to fire at them, and they began to concentrate their fire at us. Shells began to burst in the trees around us. This would shower us with red-hot shrapnel. It was a miracle that neither of us was hit! Our loaders placed high-explosive rounds into our bazooka as we waited for the signal from our loader that we were loaded and ready to fire. The noise was deafening as artillery shells exploded all around us; several German infantrymen were firing their weapons at us. How could we survive this?

BELOW This Flak 18 served in the German Condor Legion during the Spanish Civil War, and it now sits on public display in Madrid. The image presents a good view of the lower section of the carriage. *(U-95)*

ABOVE Two forms of the 88 side by side. A 8.8cm Flak 37 gun is towed past an *Elefant* tank-destroyer, the latter armed with an 8.8cm Pak 43/2. *(AirSeaLand Photos/ Cody Images)*

My weapon was loaded first as my loader tapped me on the helmet as a signal to me that he had loaded the bazooka and I was ready to fire. I got off my first round, as Private Jones was not loaded yet. After I pulled the trigger I could see my round as it left the tube and went about thirty yards before it spiraled out of control to the ground. It was a dud round!

Private Jones fired his round next and it was a direct hit on the protective shield of the 88, killing or wounding most of the gun crew. The few survivors who were left abandoned their gun and ran away.

The two remaining German 88s fired on us with a renewed determination. Shells were bursting all around us. We were being slaughtered! Realizing that to remain where we were would be certain death, Captain Wilde ordered us to charge the ridge ahead of us.

This passage offers several insights into the function and perception of the 8.8cm Flak. (Throughout this book I generally refer to the '8.8cm Flak', opting to give the calibre in centimetres as was the German preference. In US and British references, however, the gun is typically known as the '88mm' or simply as the '88'.) The first is the obvious psychological impact the presence of the guns has on the American troops; the narrator is not just alarmed at the presence of guns in general, but is specifically 'terrified' by the fact that they are '88s', known to be 'powerful'. Furthermore, this account illustrates that the fear was to a large measure well-founded; four of the M10 tank-destroyers were obliterated in the very first salvoes of fire.

The 8.8cm Flak mainly built its reputation among the Allies and the public as an anti-tank gun. In this book, however, we will also focus squarely on the role for which it was designed – delivering high-altitude anti-aircraft fire. Indeed, it was the 8.8cm Flak guns that provided the true backbone of air defence for the Third Reich. That the 88 was also a fine anti-tank gun proved to be, in part, its curse, as it was overused and sent to the most contested parts of the battlefield, often to conduct a pointless defence that simply ended with the destruction of both gun and crew. One point for sure can be said of the 88: wherever the Germans fought on land, the 88s would usually be in the front line.

Chapter One

Development

By the end of the First World War in 1918, air power was already writing the future of warfare. The 8.8cm Flak gun was born from the critical need to meet that threat with more than just truck-mounted improvisations.

OPPOSITE A restored Flak 18 ready for movement. The Flak 18's defining component is the single-piece barrel, as opposed to the stepped multi-section RA 9 barrel of the later Flak 36 and Flak 37 variants. *(LandFox/Shutterstock)*

ABOVE A German gunner sits for a reflective portrait on a Flak 18 gun at maximum elevation; the Flak 18 can be distinguished from the subsequent variants by its one-piece, unstepped barrel. *(Petjaveliki)*

Development of AA artillery in Germany began surprisingly early – actually back in the 1870s. During the Franco-Prussian War (1870–71), French forces besieged in Paris used hot-air balloons to transport mail across the battle lines, and also as an aerial platform for the launch of homing pigeons and for lofty reconnaissance. In response, the German military developed the *Ballonabwehrkanone* (balloon-defence gun), a 3.6cm breech-loading cannon bolted to the flatbed of a handcart, and with a swivel mount that allowed a high angle of elevation and a full 360-degree traverse.

Development of anti-balloon cannon continued in Germany after the end of the Franco-Prussian War – albeit in a limited way – the main players being the industrial concerns of Krupp and Ehrhardt. (Ehrhardt's founder, Heinrich Ehrhardt, established the factory for Rheinmetall – Rheinische Metallwaaren- und Maschinenfabrik Aktiengesellschaft – in Düsseldorf in 1889, and later the Ehrhardt company would be absorbed by Rheinmetall-Borsig.) Between about 1909 and 1914, Krupp and Ehrhardt produced a series of truck-mounted or horse-drawn anti-balloon guns (Ehrhardt also created a gun mounted on a field carriage), varying in calibre between 5cm and 7.7cm. All were characterised by high-angle, wide traverse mounts, while the vehicular platforms gave the weapon systems the mobility required for rapid deployment to battlefield hotspots. Significantly, the Krupp 7.5cm L/35 was designated as *Flugzeugabwehrkanone* (aircraft defence cannon), which when abbreviated gave us the 'Flak' name.

Although the development of anti-balloon cannon laid some conceptual and practical foundations for future AA artillery, it was the First World War and the interwar years that provided the necessary motivation to reach the Flak 18. During the war, aircraft rather than balloons became the key aerial threat. At first, the German AA response was rather improvisational, with captured French and Russian field guns pressed into service through various adaptive mounts and, in some instances, barrel reborings. Rheinmetall also produced the 7.7cm *Leichte* Kw-Flak L/27, formed from a 7.7cm L/27 gun mounted to a Mercedes truck.

The first steps towards the Flak 18, however, were taken in mid-1915, when both Rheinmetall and Krupp were commissioned to develop two new AA guns in heavier calibres, specifically 8.8cm and 10.5cm. The 8.8cm calibre held particular promise, offering the right blend of high muzzle velocity – more than 750m/sec (2,460ft/sec), operational convenience (the loader could handle a 15kg/33lb fixed-round shell on his own without mechanical assistance) and a successful track record within the German *Kaiserliche Marine* (Imperial Navy), hence the design process focused heavily on adapting the 1913-pattern L/45 naval gun.

The weapons that emerged in early 1916 were similar in many ways, and even a cursory glance immediately shows the visual contours of the Flak 18 taking shape. Both 8.8cm Flak

RIGHT **The Western Front, 1940. An 8.8cm Flak 18 gun crew blur in action as they engage aerial targets, firing from an emplaced position.**
(AirSeaLand Photos/Cody Images)

16 guns – the K-Zugflak L/45 Krupp and K-Zugflak L/45 Rheinmetall – had 45-calibre barrels delivering muzzle velocities of about 760m/sec (2,493ft/sec). They were mounted on twin-axle trailers, designed to be pulled by horse or vehicle, and fitted with folding outrigger stabilisers to give a solid wide-area firing platform when the gun was in action. Both guns gave respectable performances during trials conducted by the German War Ministry, and were put into production for the remainder of the war, giving the opportunity to incorporate improvements to sights, carriages and fire control, including the facility to link optical rangefinders for centralised fire control. As with so many other projects, however, production and service ceased with the final German defeat in November 1918.

The evolution of German AA artillery now had to swim through the sluggish waters of the Versailles Treaty, which placed German military development in a vice of restrictions and limitations. To circumvent the international rules, from 1921 Krupp and Rheinmetall forged foreign commercial alliances that allowed R & D to proceed in neutral countries, undetected by (or at least impervious to) the Treaty monitors. Krupp made its alliance with Svenska Aktiebolagst Bofors of Sweden, who by 1922 also had a Berlin-based office supported by the War Office, known as Koch und Kienzle. Rheinmetall went further and established its own company in Switzerland – Waffenfabrik Solothurn AG.

By such sleight of hand, both Krupp and Rheinmetall managed to remain active in all fields of artillery development during the pre-Nazi era. Indeed, it was in 1931 that Krupp/

RIGHT **An Ehrhardt 5cm *Ballonabwehrkanone* (BAK) of 1906, one of the earliest attempts to produce a mobile anti-aircraft platform, within an almost wholly armoured structure.**
(AirSeaLand Photos/Cody Images)

ABOVE The 8.8cm Flak 16 was a First World War forerunner of the Flak 18; here a captured example is seen in the hands of the Canadians. Note the use of pistons on the carriage to elevate the gun. *(AirSeaLand Photos/Cody Images)*

Koch und Kienzle unveiled the first of the guns that are the theme of this book.

Flak 18, Flak 36 and Flak 37

During the 1920s, the Krupp/Bofors association yielded several designs on paper or in prototype/limited production, one of which was the L/60, a 60-calibre 7.5cm AA gun. The L/60 entered service with the Swedish Army in 1929, and was also exported to Brazil as a 76.2mm variant to Finland and the Soviet Union. The L/60 went on to have a service career throughout the 1930s and the Second World War, and its export markets broadened to include Hungary (its biggest user), Spain, Greece and Thailand. Very limited numbers went to the German forces for dockyard air defence, but in general the ordnance authorities did not buy into the capabilities of the 7.5cm shell, looking for something with a better muzzle velocity and effective range/ceiling. In the early 1930s, by which time Krupp was once again largely operating openly and independently back in Germany, the design was revisited by Krupp engineers, this time with a view to incorporating the 8.8cm shell. The new gun calibre and design was prompted by the need to reach the next generations of monoplane bomber aircraft, which took operational altitude up to and beyond the 7,620m (25,000ft) mark.

It was this gun, unveiled in prototype form in 1932, that brought us the Flak 18 L/56. The Flak 18 was an 8.8cm AA gun that delivered impressive firepower for the time, even when compared to heavier-calibre models, such as 10.5cm guns. With a competent crew, it

RIGHT A well-restored Flak 37 gun. This photograph was taken at standing human height, giving a good impression of the height profile of the gun on its wheels. *(Author's collection/Axis Track Services)*

RIGHT This 8.8cm Flak, part of the Muckleburgh Collection, saw service in the Spanish Civil War. It is a hybrid gun, with a Flak 18 barrel but upgraded with the *Übertragungsgerät 37* receiver system of the Flak 37.
(Author/Muckleburgh Collection)

could fire high-explosive (HE) shells with a muzzle velocity of 820m/sec (2,690ft/sec), reaching up to an effective ceiling of around 8,000m (26,250ft). The gun had a fast rate of fire of 15–20 rounds per minute, by virtue of its semi-automatic mechanism, which ejected the spent shell cases smartly and re-cocked the gun for the next round during the recoil phase; reloading was via either hand or by a power-operated rammer. The gun was mounted on a cruciform platform that when deployed allowed the gun to make two full 360-degree turns on

BELOW A weathered Flak 36/37 at Aberdeen Proving Ground in Maryland, USA. The drums at the front would hold the communications cables that linked the weapon to the fire-control director.
(Mark Pellegrini)

DEVELOPMENT

ABOVE Another Flak 36/37 at Aberdeen Proving Ground. Unlike its companion, this one is set on a dual-wheel *Sonderanhänger 202*. *(Mark Pellegrini)*

RIGHT In this pre-war drill, conducted at one of the Nazi Party's incessant propaganda rallies, an 8.8cm Flak 18 shows its application for anti-aircraft fire. The man in the foreground unpacks 8.8cm shells. *(AirSeaLand Photos/ Cody Images)*

a stable firing base. For transportation, the gun and the platform were towed on a two-axle carriage behind a semi-tracked vehicle, typically an SdKfz 7 half-track. It could be taken from its carriage and installed ready to fire in just 2 minutes and 30 seconds, and fire control was via a gun control telescopic sight – the *Flakzielferhrohr 20* or the *Übertragungsgerät 30* predictor firing system. The Flak 18 was a convincing package, and it went into service with the German Army in 1936. (Incidentally, it was called the Flak 18 to confuse the Treaty monitors about when it had entered service.)

Yet a problem emerged, for which Rheinmetall found a rather over-engineered solution. The high rate of fire of the Flak 18, plus the copper driving bands fitted to the HE shells of the time, meant that the gun's monobloc barrel (i.e. produced as a single unit) only had a service life of about 900 rounds. Experience from the gun's use in the Spanish Civil War (1936–39) quickly illustrated that this barrel life would not be sufficient for high-tempo combat conditions. The Rheinmetall solution was the *Rohr Aufbau 9* (RA 9) multi-section barrel. This barrel consisted of five components in total: (1) an outer jacket; (2) an inner sleeve; (3–5) a three-section inner tube, holding the rifled portions of the barrel. The thinking behind the RA 9 was that the greatest wear to the barrel took place at the forcing cone (the section in which the shell transitions from the chamber to the bore), so when this wore out the RA 9 modular structure meant that just that section could be replaced, rather than the whole barrel. (More about barrels, indeed all technical aspects of the Flak series, can be found in Chapter 2.)

The introduction of the RA 9 to the Flak gun in 1937 brought with it the first variant of the series, the Flak 36, visually distinguished from its predecessor by the stepped outer profile of the new barrel, caused by the barrel jacket locking collar two-thirds of the way along the length. And yet the RA 9 was not destined to be the ideal solution to the barrel wear problem, especially under war conditions. The development of propellants with lower burn temperatures and improvements in ammunition (particularly the introduction of sintered-iron driving bands rather than

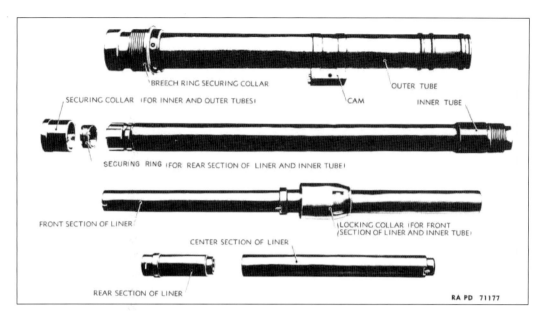

LEFT A US military diagram of the RA 9 composite bore for the Flak 36/37. Note the 'Rear Section of Liner', the part that contained the easily worn forcing cone. *(US War Department)*

RIGHT A diagram of the Flak 36 set on its cruciform platform, the carriage levelled through the use of the levelling jacks at the end of the outriggers. *(US War Department)*

bore-punishing copper ones) meant that the barrel life could be extended anyway. Furthermore, the RA 9 required intensive precision engineering to manufacture, in order to ensure the fine tolerances between each element. The production expense might have been acceptable in peacetime conditions, but under the huge pressures on manufacturing capacity during a world war, it added a further burden. In addition, in hard field use the RA 9 sometimes suffered from its very precision, the joints between components swelling or being contaminated by debris. Such is not to say that the RA 9 was a failure – thousands of these barrels destroyed great numbers of tanks and aircraft during the Second World War. But it speaks volumes that eventually the RA 9 was simplified to a two-section inner liner, then in the last years of the war the manufacturers returned to monobloc construction.

RIGHT Flak 18 guns sit awaiting transit on their twin-bogie *Sonderanhänger 201* wheeled carriage, identifiable by the single set of wheels on one bogie and the facing-travel orientation of the gun. *(AirSeaLand Photos/Cody Images)*

RIGHT In this posed photograph of the Flak 18 in action, a shell is fed into the breech from the loading tray while the two men on the right watch their traverse and elevation receivers intently. *(AirSeaLand Photos/ Cody Images)*

FAR RIGHT In this scene from Budapest in the last days of the Second World War, an 8.8cm Flak has been installed in a crudely armoured Vomag truck. Such adaptations tended to be unsuccessful, creating an easier target for Allied anti-tank teams. *(AirSeaLand Photos/ Cody Images)*

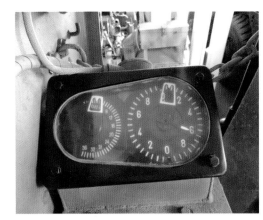

RIGHT The *Übertragungsgerät 37* fire-control receiver system, this 'Receiver C' component being for the fuze-setter, providing instruction on the correct shell fuze setting for the required detonation altitude. *(Author/ Muckleburgh Collection)*

The Spanish Civil War proved useful to the German military renaissance of the 1930s in many ways, providing some early battlefield testing of fresh tactical thinking and new technologies, particularly with regard to aviation, armour and artillery. The Flak 18 was heavily deployed as part of the Condor Legion. Although it served in its AA role, the Flak 18 also showed its early potential as an anti-tank (AT) gun, something that the Germans had sensed from the outset. Field reports and other feedback flowed back through the German military command and on to the factory floor at Krupp and Rheinmetall, resulting in some upgrades. Many of the changes focused on the wheeled carriage. The original *Sonderanhänger 201* was updated in the form of the *Sonderanhänger 202* by 1939. Of several changes associated with the limber, the most important was that the gun could now be towed with the barrel pointing in either direction (having the barrel pointing to the rear meant that it could be brought into action with greater speed). In addition to this, the fitting of twin wheels on each bogie, combined with other improvements in robustness, meant that now the 8.8cm Flak could be fired direct from its wheels if need be. Firing from the wheels was not ideal – the gun's recoil was punishing for all limber components and accuracy was compromised – but it gave the crew the reassurance of an extra-fast deployment time to meet sudden threats.

The next significant change was in relation to aerial fire control. Simply described, on the *Übertragungsgerät 30* system, the information from the predictor and rangefinder was transmitted to gun-mounted receiver units that consisted of three concentric rings of electric bulbs and three mechanical pointers, each pointer of a different length. Gunlaying

ABOVE The twin-wheeled bogies of the *Sonderanhänger 202*, detached and displayed side by side. Each wheel had its own connection to the braking system. *(Yaco Erisso)*

LEFT The Flak 37 was defined by the change to the *Übertragungsgerät 37* fire-control receiver system, the twin receivers of which are visible here. Also note the two ports for connecting headsets when receiving radio fire-control instructions. *(D.W.)*

ABOVE This Flak 37 is on display a long way from home, in the Cavalry Tank Museum, Maharashtra, India. Indian Commonwealth forces encountered the 8.8cm guns in North Africa and Italy.
(Mohit S)

combined with the *Sonderanhänger 202* and the updated RA 9 with two-piece inner barrel liner together formed the Flak 37. (Note also that old monobloc barrels were also used on the Flak 36 and Flak 37; indeed all barrel configurations were interchangeable between types.) The next steps in the Flak 88's evolution, however, would be conducted under the chaos and destruction of a world war.

The Flak 41

The onset of war in 1939 meant that the 8.8cm Flak guns were rushed into action and accelerated production. They quickly began to build a grim reputation for their fearsome abilities to kill both tanks and aircraft, especially during the battles in France in 1940 and in the Western Desert and North Africa in 1941–42. For many on the Allied side, it was the 88's tank-killing role that took the psychological headlines, probably for the simple reason that flying 5 miles up it is hard to tell quite what is shooting at you, whereas at ground level the enemy artillery could be identified with a pair of binoculars. Yet even as the war began the Luftwaffe (the principal arm in charge of AA warfare) was looking to make improvements on the gun that would keep pace with the evolution of bomber aircraft. For example, the B-17G

information was fed to the receiver units and the information showed up via lit bulbs; the gunners then operated the traverse and elevation handles to cover the bulbs with the tips of the pointers. The system worked, but proved to be somewhat cumbersome, and in 1939 its replacement – the *Übertragungsgerät 37* – entered production. In this version, the receiver dials featured pointers that were electrically controlled by data feeding through from the predictor; all the gunners had to do was turn the elevation and traverse wheels until the gunners' pointers matched the others. The system was reassuringly quicker and more reliable. The *Übertragungsgerät 37*,

RIGHT A Flak 37 on display in the Museo Histórico Militar de Cartagena, Spain. The black-painted bell on the side was used to provide an auditory fire signal when necessary.
(Yaco Erisso)

variant of the Boeing B-17 Flying Fortress was introduced in 1943 with a service ceiling of 10,850m (35,600ft). Although in 1939 such operational altitudes were not yet in evidence in production aircraft, the Luftwaffe knew the writing was on the wall, and wanted its AA guns to match the threat.

Thus a specification was issued, named *Gerät 37*, for a gun that could fire 25 rounds per minute at a muzzle velocity of 1,000m/sec (3,281ft/sec), the whole gun weighing no more than 8,000kg (17,637lb). Rheinmetall tackled the challenge, naming the project the 8.8cm Flak 41. In terms of its engineering ingenuity, the Flak 41 that emerged from the Rheinmetall factory was undoubtedly brilliant. It achieved all the *Gerät 37* specification requirements. To give it the 25 rounds per minute performance, it included an auxiliary hydropneumatic roller loader mechanism, in which a piston was withdrawn and held in the cocked position during the recoil phase of firing, then used stored air pressure to ram a shell home via rubber gripper rollers as soon as the loader placed a shell in the loading tray. The gun had three electrical firing circuits, for air defence, ground defence and emergency firing. It was also placed on a turntable mount that not only gave the gun a very low profile – always welcomed by crews trying to make their gun emplacement as inconspicuous as possible – but by having the trunnions at the very rear of the carriage it meant the gun could be elevated to a full 90-degree angle upwards.

The one part of the Flak 41 that really let it down, however, was its barrel. Like the RA 9, the Flak 41 barrel was a multi-section affair, and a complicated one at that. The principle of sectional barrels was sound, but when delivering intensive fire, and especially when firing the cheaper steel-cased (as opposed to the more expensive brass-cased) ammunition types, some serious problems emerged as the barrel sections heated up and adjusted their dimensions. Gun historian Ian V. Hogg explains the problem, and the hunt for the solutions:[2]

> *This was all very well on paper, but in practice it gave a good deal of trouble owing to the cartridge case failing to extract. Part of the trouble was due to the fact that the joint between the first and second sections of the barrel fell exactly at the cartridge case bottleneck; steel cases expanded into this joint and became stuck. Special brass cases were developed, which overcame the trouble to a large degree, but it periodically recurred; this was probably due to the high chamber pressures involved. The design was eventually changed to a two-section liner*

BELOW A rare surviving Flak 41 gun, a powerful weapon here fitted with a full gun shield for the crew. The barrels were produced in five-, four- or three-section versions, depending on the year of manufacture. *(Mark Pellegrini)*

ABOVE Another angle on the Flak 41. Although the shield indicates this gun was intended for front-line combat use (some saw action in North Africa), most ended up in static home defence AA positions. *(Mark Pellegrini)*

with a jacket and a sleeve. The 152 guns issued with the original three-piece liner were marked with a yellow band around the barrel and the yellow M painted on the breech-ring, indicating that they were only to be used with ammunition having brass (Messing) cartridge cases. A further 133 guns were then issued with two-piece barrels, but the trouble persisted – though to a smaller degree – and was largely due to the different expansion factors of the chamber and the cartridge case when under severe pressure. The design was again changed to a heavier two-piece barrel with a jacket and no sleeve, and the remainder of production (some 271 guns) used the third type of barrel.

Hogg's description of the efforts to undo the problems of engineering sophistication could be applied in general principle to many German weapon projects, which often tended towards brilliance rather than simplicity, and therefore struggled in the base dirt and heat of war. Yet Hogg also points out that the Flak 41 was still 'a good weapon'. Drawing on its 8.8cm Flak ancestry, the Flak 41, when it was working properly, delivered shells far, fast and accurately, a nightmare for an enemy high-flying bomber crew or tank force.

The Flak 41 went into production in the spring of 1942, but straight away the *Heereswaffenamt* (German Army Weapons Agency) faced the issue that the volumes produced would never meet the sizeable and growing demand from the field. Therefore, the option was explored for taking the new Flak 41 gun and mounting it on the simpler Flak 18/36/37 carriages. The idea was a non-starter, largely because the high stresses imparted by firing the Flak 41 were greater than the Flak 18/36/37 carriages could sustain.

This was by no means the end of the experiments, however, as the Germans sought other routes to achieve the required high-velocity weapon. One avenue of exploration was simply altering the barrel length of existing Flak guns to give them a greater muzzle velocity, the extra recoil controlled to varying degrees of success by muzzle brakes. One of these was a hybrid weapon, the Flak 37/41. From the Flak 41 came the auto-loading mechanism plus the gun's fuze-setting mechanism. The barrel was 6.54m (21ft 5.5in) – by comparison, the length of the Flak 18/36/37 was 4.93m (16ft 2in), but the gun retained the Flak 37's *Übertragungsgerät 37* fire-control system. The Flak 37/41 might have had promise on paper, but again it was blighted by overcomplication.

Not only did the hybrid nature of the gun add to Germany's already white-hot manufacturing demands, but it also retained the Flak 41's problems regarding shell extraction. On account of these factors, very few Flak 37/41 units were produced, probably fewer than 20 in total by the war's end.

The desire to increase the performance of the pre-war generation of 8.8cm Flak guns not only yielded the Flak 41. In 1941, Krupp was also approached to take the Flak firepower to even greater heights (literally), with the *Gerät 42* specification. Here the muzzle velocity was increased to 1,020m/sec (3,608ft/sec) with a heavier shell weight – 10kg (22lb) – meaning that more explosive weight could be hoisted to greater altitudes. Krupp decided to take a family approach to the new gun, meeting the specification with a single weapon that could be mounted as an AA gun, a dedicated AT gun and a tank main gun. Krupp was making progress in this work when, in late 1942, the Luftwaffe arm responsible for AA artillery again altered the specification, taking the shell weight down to 9.4kg (20lb 11.5oz) but increasing the muzzle velocity to 1,100m/sec (3,609ft/sec). This sudden interruption of development essentially killed the Krupp programme, and the Flak component of the *Gerät 42* specification was cancelled. The Flak 41 would essentially be the last in the line of the Flak AA guns.

Diversity

By the end of the Second World War, roughly 20,750 Flak 18/36/37 guns had been manufactured, by far the most numerous of the 8.8cm Flak series; the Flak 41s numbered only about 544. As we shall see in later chapters, the 8.8cm Flak guns made a bloody contribution in every theatre to which they were deployed. In some actions a single 8.8cm crew might kill double-figure numbers of enemy tanks, hitting home every three or four shots. It was a gun that certainly exerted influence.

ABOVE This excellent side view of an 8.8cm Flak 37 shows the gunlayer's seat, the elevation and traverse wheels, the mount for the *Zielfernrohr* (telescopic sight) and the twin receiver dials of the *Übertragungsgerät 37*. *(AirSeaLand Photos/Cody Images)*

RIGHT A Flak 37 lies in the shattered ruins of Berlin in 1945, its barrel still angled upwards at the Allied airpower it found impossible to defeat. *(AirSeaLand Photos/Cody Images)*

ABOVE This Flak 18, on display in the Netherlands, was part of the Atlantic Wall coastal defences, emplaced both for anti-aircraft work and for fire against offshore naval traffic or amphibious forces. *(Paul Hermans)*

ABOVE RIGHT Shown alongside a searchlight, this Flak 37 is in full elevation; to elevate from −3 to +85 degrees took 15.02 seconds at maximum speed (with the high gear engaged). *(Pappenheim)*

BELOW German Pak 43/41 anti-tank guns, identifiable by the two-wheel split-trail carriage from the 10cm le K 41 (10cm *Leichte Kanone 41*), lie abandoned on the Eastern Front, along with their SdKfz 251 prime movers. *(AirSeaLand Photos/Cody Images)*

The principal focus of this book is on the technicalities and operations of these specific weapons in particular to give our narrative clarity, technical focus and tactical insight. Yet it would be a dereliction of duty not to touch upon migration of the 8.8cm gun into other mounts and contexts. Doing so rounds off our understanding of the potential and capability of this extraordinary artillery piece.

Anti-tank guns

The Flak 18/36/37 were all, primarily, AA guns, but whose excellent ballistic capabilities – when paired with the right sighting, fire control and ammunition – made them formidable and practical AT guns. But the anti-armour properties of the 8.8cm gun made it almost inevitable that a dedicated AT weapon would emerge from the Flak development programme. In this case, the point of intersection was the aforementioned Krupp investment in the *Gerät 42* specification. As noted above, the Krupp programme looked to develop a family of interrelated weapons based on the same gun type, sharing some common ammunition. While the Flak avenue of the programme was ultimately blocked, the other path led to an AT gun – the 8.8cm *Panzerabwehrkanone 43* (Pak 43) and the KwK 43 tank gun.

The Pak 43 was an exceptional weapon in all senses, and it brought a new level of threat to the battlefield when it entered service in 1943. Like the Flak guns, it was deployed on a cruciform platform, mobilised on two-wheel trailers. The platform did not need the high angle of elevation required by the anti-aircraft guns, therefore the gun could hunker down closer to the ground and presented an extremely low visual silhouette – just 1.73m (5ft 9in), roughly the height of an average human. Dropped down into an emplacement, little more than half of that height might be above ground. If need be, however, the gun could also be fired from its wheels.

The Pak 43 performance had similar characteristics to the AA guns, hurling out a 10.4kg (22.93lb) armour-piercing shot at 1,000m/sec (3,281ft/sec), giving an armour penetration (0 degrees) of 207mm (8.15in) at 500m, 190mm (7.48in) at 1,000m and 174mm (6.85in) at 1,500m). A key difference between the operating mechanism of the Flak guns and the Pak 43 was the breech mechanism, which used a vertical sliding block. The breech mechanism also included a dual-spring mechanism, loaded under tension during recoil, that powered breech opening and ejection during the recoil stroke, and breech closing prior to firing the next round. This system made the gun very quick to operate, with a short recoil stroke and a snapping efficiency of movement.

The Pak 43 was arguably the best anti-tank gun of the war. An attempt to make production catch up with demand led to the Pak 43/41, a less-than-effective mix of the Pak 41 barrel married to breech mechanisms, carriages, wheels and bits and pieces from other guns, including a split trail from a 10.5cm le FH 18. This weapon was still to be treated with deference by enemy armour crews, but it was an awkward and mechanically problematic version that was not respected by AT gun teams.

Tank/SP guns

The performance characteristics of the 8.8cm gun made it naturally attractive to the designers of German armoured vehicles. Although Germany began the Second World War with some decent enough tanks and self-propelled (SP) guns, it soon became apparent (particularly with the appearance of the T-34 tank following the launch of Operation Barbarossa in June

1941) that both armour and gunnery needed to keep pace with developments. This is why the 8.8cm Flak guns were fitted to various armoured mounts. Many of these conversions were just experiments, resulting in nothing more plentiful than a handful of prototypes, or at best very limited production figures. Some of them, however, became genuine forces of influence on the battlefield.

The first tank to receive the 8.8cm Flak was actually one of the landmark German combat vehicles of the war – the PzKpfw Tiger Ausf E, SdKfz 181. The Tiger I (as it is better known) entered production in August 1942, and it was

ABOVE A good view of the breech end of a (captured) Pak 43 anti-tank gun, with one of its armour-piercing rounds in the foreground. Note the gunner's seat on the left, with the *Zielenrichtung 43 v so* direct-fire sight looking through the shield. *(AirSeaLand Photos/Cody Images)*

LEFT The muzzle end of the Flak 18 barrel, clearly showing the right-hand twist rifling, which by the muzzle end was 1 turn in 30 calibres. *(Author/Muckleburgh Collection)*

ABOVE A close-up of the breech of an 8.8cm KwK 36 gun, the firepower centrepiece of the Tiger I tank. Note the vertical configuration of the sliding breech-block, not horizontal as in the 8.8cm Flak weapons. *(AirSeaLand Photos/Cody Images)*

a monster compared to all other armoured vehicles. Weighing in at 54 tonnes (60 short tons), most of that weight was accounted for by its prodigious size – total length with gun forward was 8.45m (27ft 9in) – and its extremely thick armour, which went up to 120mm (4.7in) on the gun mantlet. Armament came in the form of the 8.8cm KwK 36 gun. We must exercise some caution about drawing a direct line between the Flak 36 gun and the KwK 36. It is probably safest to say that the KwK 36 was inspired by the AA weapon, and then proceeded on an independent but related line of development. Both shared a 56-calibre barrel length and ballistic properties – the Flak guns had already proven that they were tank killers, so the wheel was not reinvented unnecessarily. The key differences between the two were that KwK 36 had a different muzzle brake, breech system (a falling wedge type rather than a sliding block), electrical primer ignition, a one-piece barrel rather than multi-piece barrel and also different sighting arrangements suited to an armoured vehicle. Despite the physical differences, the ancestral connection between the Flak 36 and the KwK 36 was still apparent.

In 1943, the Germans also unveiled the Tiger II, more commonly termed the *Königstiger* (Royal Tiger, or less accurately by the Allies, King Tiger). The Tiger II tipped the scales even further, taking the overall weight up to an impractical 68.5 tonnes (75.5 short tons). To deliver even greater destructive punch than its predecessor, the Tiger II also used an 8.8cm gun, but this version was the 71-calibre KwK 43 used in the Pak 43. This weapon gave the Tiger II the same killing power as the AT gun, but with an onboard store of 80 rounds (typically split between 40 HE rounds and 40 armour-piercing (AP) rounds), all wrapped in fortress-like armour. Yet the Tiger II's dimensions, and the sheer cost and effort involved in its production, were simply not worth it for the overall return on the battlefield. Most Tiger IIs were either destroyed on their production lines by Allied bombing or suffered from mechanical breakdown. The tank also shared the painfully slow turret traverse of its predecessor, which often meant that targets slipped out of view of the formidable gun.

Better use of the 8.8cm guns was made by utilising them in SP tank-destroyers. While such vehicles did not have the survivability of the big tanks, they were more cost-effective enemy armour-killers than hefty beasts such as the Tiger. The first SP guns designed specifically

LEFT A Canadian infantryman in Italy proudly stands atop the *Nashorn* tank-destroyer he knocked out, his PIAT anti-tank weapon on his back and an ammunition belt draped over the muzzle brake of the 8.8cm Pak 43/1. *(AirSeaLand Photos/Cody Images)*

around 8.8cm Flak models were the *Flak 18(Sf) auf Zugkraftwagen 12t*, SdKfz 8, and the *Flak 37 Selbsfahrlafette auf 18t Zugkraftwagen*. The names of these vehicles indicate the source of their armament; the key difference between the two were the chassis on which they were mounted – the Flak 18 went on a Daimler-Benz DB10 ZgKw 12-tonne half-track tractor, while the Flak 37 was mounted on an SdKfz 9 half-track. Armoured protection was provided for the gun crew on the back of the vehicle.

Neither of these vehicles was produced in large numbers, although handfuls saw limited active service in France and Italy.

Other SP versions of the '88' soon followed, including the experimental VFW 8.8cm *Flak auf Sonderfahrgestell* (Pz Sfl IVC) – a half-track-mounted L/71 gun intended for mobile AA work. But the first really successful SP 8.8cm was to the SdKfz *Hornisse* or (as it was renamed in 1944) *Nashorn* tank-destroyer. Designed between October 1942 and May 1943, when it went into production, the *Hornisse* merged a Geschützwagen III/IV chassis with an 8.8cm Pak 43/1 L/71, and was in many ways a more potent version of the 15cm *Schwere Panzerhaubite Hummel*. The gun was fixed to the rear of the chassis, with the open-topped fighting compartment wrapped in armour that measured 15mm (0.59in) at the front and 10mm (0.39in) at the side. The gun was also set in a 15mm armour mantlet, the fitting giving the weapon a 15-degree traverse to each side and an elevation of 5–20 degrees.

The *Hornisse/Nashorn* was a well-made and tactically useful weapon that more than proved its worth on the Eastern Front, capably tackling some of the new generations of Soviet heavy tanks such as the KV-1 at long ranges. It also had the virtue of being cheap to manufacture, at least in comparison to a main battle tank.

The 8.8cm gun was applied to two other SP tank-destroyers. The first was the SdKfz 184 *Ferdinand/Elefant*. This vehicle was actually a solution to utilising 100 chassis created during Ferdinand Porsche's failed attempt to win the contract for the Tiger tank design. With the German forces impressed by the performance of the *Hornisse/Nashorn*, it was decided to turn 91 of these chassis into very substantial tank-destroyers armed with the Pak 43/2 L/71 gun. Supplied with 50 onboard rounds, the *Elefant*'s gun had a

ABOVE In this fearsome array of 8.8cm firepower, a formation of Tiger IIs prepare for deployment to Hungary in 1945. The capabilities of the gun were let down by the vehicle's lack of reliability. (AirSeaLand Photos/Cody Images)

RIGHT Only 12 8.8 cm Flak 18 (Sfl) auf Zugkraftwagen 12t (SdKfz 8) were produced prior to the outbreak of war in 1939. Traverse was very limited to either side of the vehicle, compromising the versatility of the 8.8cm Flak gun. (AirSeaLand Photos/Cody Images)

ABOVE The *Elefant* was a thunderous tank-destroyer, equipped with the 8.8cm Pak 43/2, a 71-calibre gun that could kill any Allied armoured vehicle present on the battlefield. *(AirSeaLand Photos/Cody Images)*

RIGHT One of several attempts to make a cheap tank-destroyer – the 8.8cm Flak 18 (Sfl) auf Zugkraftwagen 12t (SdKfz 8). Note the limited space for the gun crew atop the vehicle. *(AirSeaLand Photos/Cody Images)*

RIGHT The *Jagdpanther* was armed with the Pak 43/3. Up to 60 rounds of 8.8cm ammunition could be stored within the armoured fighting compartment. *(AirSeaLand Photos/Cody Images)*

traverse of 30 degrees, and an elevation and depression of –8 to +8 degrees. Yet, as with the Tiger tank, the *Elefant* was formidable – it had front armour of maximum 200mm (7.87in) thickness – but its sheer bulk was its undoing, being vulnerable to mechanical breakdown, running out of fuel and being outmanoeuvred by far more nimble Allied opponents.

By contrast, the *Jagdpanther* (Hunting Panther) delivered far better results. This vehicle took the Ausf G chassis of the superb PzKpfw V Panther tank and added an armoured superstructure and a Pak 43/3 L/71. Being a late-war vehicle – production began in January 1944 – the *Jagdpanther* appeared to have learned some further lessons about reliability, and had an improved Maybach HL 230 P30 V-12 petrol engine and a stronger transmission when compared to the standard tank, meaning that it remained dependable even under hard use. As with all SP tank-destroyers, the traverse (3 degrees left/right) and elevation (14–80 degrees) were limited, but with nimble positioning by the driver the *Jagdpanther* was taking accurate shots out to 3,000m (3,281yd).

The final SP mount for the 8.8cm Flak/Pak gun was the *Panzerjäger* Tiger Ausf. B, or SdKfz 186 *Jagdtiger* (Hunting Tiger). A late, and frankly pointless, addition to the German arsenal in 1944–45, the *Jagdtiger* was actually the heaviest tank of the war, at 71.7 tonnes (79 short tons). It was intended to carry a 12.8cm Pak 44 L/55 main gun, but as not all were available for even the small number of *Jagdtigers* produced – just 70–88 – then in some instances an 8.8cm Pak 43/3 stepped in as replacement. It was a waste of a good gun. The *Jagdtiger* was a mechanical nightmare and broke down frequently. Those that did enter combat were often badly handled by inexperienced late-war crews, and many of the vehicles were destroyed under multiple tank-gun strikes or by strafing and rocket runs by ground-attack aircraft.

Other mounts

In addition to the examples outlined above, the guns and the component parts of the Flak 8.8cm weapons appeared in a range of other physical and tactical settings. Batteries of Flak 18/36/37/41 guns were mounted on specially adapted railway wagons to provide some measure of rail-mobile AA defence. Known as *EisenbahnFlak* (Flak Train), these batteries tended to consist of Flak 18/36/37 guns set on /2 static mounts. In terms of the wagon type, some were extremely crude – basically little more than a flatbed goods wagon – but in many cases they were specifically converted for artillery service. The main conversion was known as the *Geschützchwagen III (Eisb) schwere Flak*, which featured heating pipes, drop-side working platforms, ammunition lockers and car stabilisation screw jacks to support firing.

The need to project critical points on the battlefield, even aquatic ones, also resulted in 88s mounted on ferry platforms, known as the *Siebel Fähre* (Siebel Ferry), named after the system's designer, a Luftwaffenoberst Siebel. Essentially the Siebel Ferry was constructed from bridging pontoons to create a two-hull platform powered by surplus aircraft or truck engines. Each ferry was capable of taking up to three 8.8cm Flak guns (or other weapons), and could sail around at speeds of up to 9 knots. They were deployed to calm coastal areas in the Balkan/Mediterranean theatre, and were also used to provide AA defence at vital crossing points on inland waterways, such as key bridges.

As well as putting the guns themselves to good use, the Germans also utilised the carriages of the Flak series to various purposes, including rocket launching systems. Most of these applications were more experimental than practical. Turning away from such deviations, however, we now look more closely at the engineering of the main guns.

ABOVE The *EisenbahnFlak* (Flak Train) – Flak 37 guns mounted on rail flatbeds, the adaptation including anti-fragment shields and storage lockers for ammunition and gun tools. *(Bundesarchiv, Bild 101I-638-4208A-25/Hagen/CC-BY-SA 3.0)*

LEFT This German armoured train has an 8.8cm Flak gun installed for air defence. The gun has a full armour shield to protect the gun crew from small-arms attack or strafing from the air. *(AirSeaLand Photos/Cody Images)*

LEFT A rare picture of what appears to be a Flak 36 mounted to the foredeck of a Kriegsmarine coastal craft, used for air defence plus direct fire against light enemy shipping. *(Bundesarchiv, Bild N 16035/Horst, Grund/CC-BY-SA 3.0)*

Chapter Two

Design and basic operation

The 8.8cm Flak guns were the products of first-rate German engineering. Made with precision (albeit excessively so in part) and with an eye to combat efficiency, the 8.8cm Flak managed to blend form and function almost perfectly.

OPPOSITE This side view of the Flak 37 shows some of the main components of the carriage, from top to bottom: recuperator, barrel, recoil cylinder, equilibrator. *(Author/Axis Track Services)*

The barrel and breech

ABOVE This view of the breech section shows, to good effect, the horizontal sliding breech-block. The circular fitting in the base of the block is the firing spring retainer. *(Author/Muckleburgh Collection)*

The barrel of the Flak 18 was, as described earlier, of monobloc construction with a length of 56 calibres, which translated to a bore length of 470cm (185in), and an overall gun length of 493.8cm (194.1in). The rifling within the barrel had a right-hand twist, increasing from 1 turn in 45 calibres to 1 turn in 30 calibres, with a total of 32 grooves overall.

With the RA 9 barrel, the dimensions and the rifling were unchanged, but the construction was radically different. It consisted of five main parts: (1) a half-length outer tube (giving the Flak 36/37 the stepped appearance about a third of the way down the barrel); (2) a half-length inner lock tube; and (3–5) a three-piece barrel liner, containing the rifled sections of the barrel. The half-length outer tube was fitted with a breech ring securing collar; the inner tube had a securing collar and securing ring, about which more shortly. The all-important liner was, as noted, separated into three sections, with the front and centre sections containing the gun's rifling. For this reason, as noted in the US War Department's wartime manual TM E9-369A *German 88-mm Antiaircraft Gun Material*:[3]

The Flak 8.8cm (here we shall use that phrase as shorthand for the Flak 18, 36 and 37, the most common types) was the unification of four major subsystems – the gun (barrel and breech), recoil mechanism, mount and sighting/fire control. To these subsystems we can also add the method of transportation, but that will be the subject of a later chapter.

[the] front and centre sections of the liner are keyed in place so as to aline [sic] the rifling

RIGHT The highly visible locking collar on the RA 9 barrel, which prevented forward movement of the liner and inner tube, was a distinguishing feature of the Flak 36 and Flak 37 guns. *(Author/Axis Track Services)*

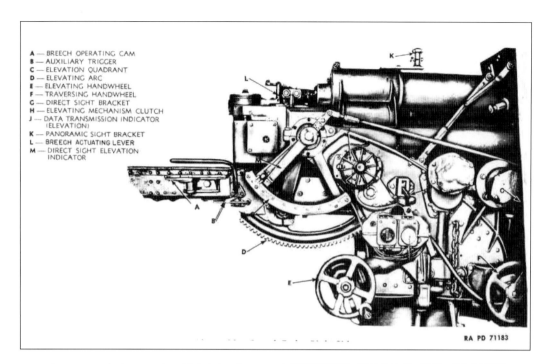

LEFT A US military diagram of the right side of a Flak 36 breech. Features of note include the *Übertragungsgerät 30* elevation receiver (J) and, atop the recuperator, the mount for the *Rundblickfernrohr 32* (Rbl. F. 32; Panoramic Telescope 32) (K). *(US War Department)*

and prevent relative rotation. This joint does not have any seal other than that provided by close tolerance machining. The centre and rear sections are merely overlapped and not keyed in place as there is no rifling to aline.

The rear section of the liner did not have any rifling because it included the unrifled chamber section of the gun. Once the three liner sections had been assembled together, the inner tube was placed over them to lock the whole structure into place with the chamber, the securing collar on the inner tube engaging with the locking collar on the front section of the liner.

Note that the three inner tubes of the barrel were not of equal lengths – making them so would partially defeat the practical and economical intentions behind making a multi-component barrel in the first place. In fact, the front (muzzle end) section of the liner extended nearly two-thirds of the length of the rifled section; the total length of the rifled section was 400cm (157.2in). The short rear section of the liner, containing the chamber and forcing cone, actually terminated about 15cm (6in) before the rifled section began. The idea was that it was

RIGHT A top view of the breech mechanism. The circular component with the attached handle is the breech actuating control. *(US War Department)*

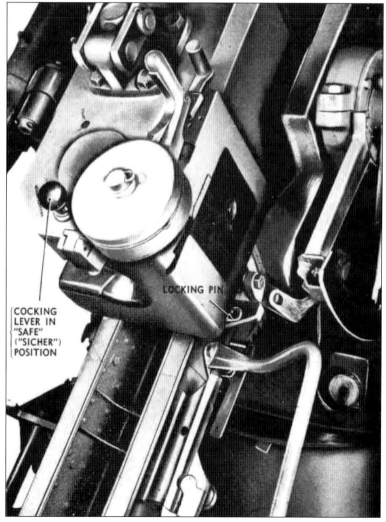

ABOVE The muzzle rest took pressure off the barrel when the gun was in transit or in storage; the handle at the side is the muzzle rest lock. *(Author/Axis Track Services)*

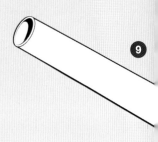

easiest to replace that frontal section of the barrel – which endured the hardest wear from firing – than it was to replace the entire barrel.

Looking at the rear of the gun, the breech mechanism was of the horizontal sliding breech-block type, meaning that a large metal slab (the breech-block) slides horizontally across to the right to expose the open chamber for loading a shell, and slides back across to the left to lock the shell inside for firing. The key advantages of the sliding breech-block design are its strength under extreme firing pressures, its simplicity and the rapidity of operation. The Flak 88 breech-block could be set for either manual or semi-automatic operation. For the former, the breech-block was closed and opened manually by the loader operating the breech actuating mechanism on top of the breech. When the gun was set for semi-automatic operation, however, the action

8.8cm Flugzeugabwehrkanone. *(Mark Rolfe)*

1 Levelling jack
2 Barrel support mechanism
3 Spring equaliser
4 Leveller wheel
5 Traverse mechanism
6 Aiming assembly
7 Barrel cradle
8 Recoil cylinder
9 Barrel
10 Barrel ring (for barrel support mechanism)
11 Recuperator
12 Breech actuating mechanism
13 Rammer
14 Loading tray
15 Upper mount
16 Elevating arc
17 Fuse adjuster mechanism
18 Fuse setting mechanism
19 Pedestal mount
20 Stabiliser outrigger
21 Outrigger stakes
22 Fuse adjusting seat
23 Leveller wheel
24 Traverse mechanism

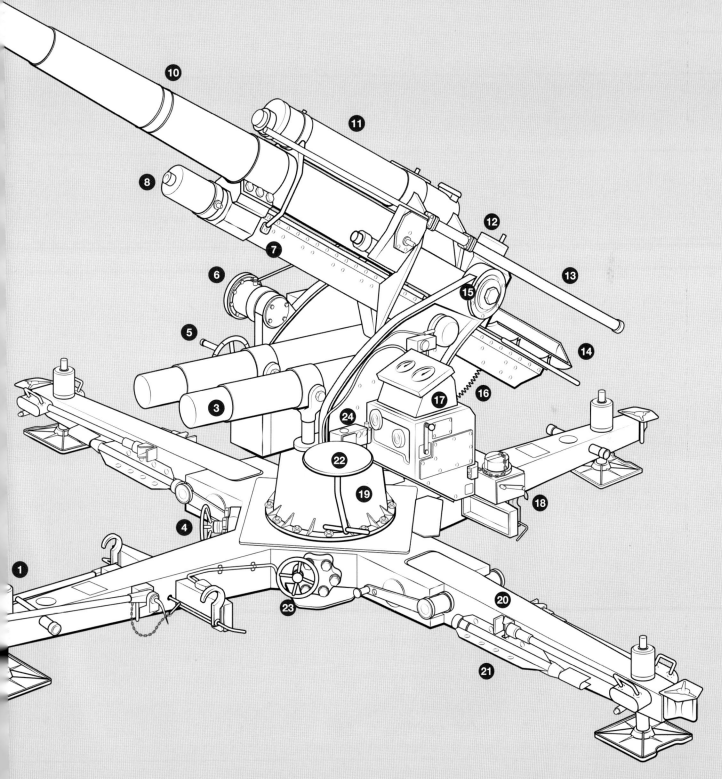

33
DESIGN AND BASIC OPERATION

of loading a round of ammunition tripped a mechanism that closed the breech mechanically under the power of a breech actuating spring. The breech was also opened automatically during the recoil phase, and the spent shell case ejected.

The Flak 8.8cm's firing mechanism was of the percussion type, meaning that a firing pin was driven against a primer in the base of the shell to start the process of propellant detonation (as opposed to primer ignition via electrical charge). As long as the cocking lever on the top of the breech ring (the housing in which the breech-block slid) was set to 'FEUER' (Fire), the firing pin was cocked automatically every time the breech-block was opened, either manually or by recoil. The cocking lever could also be used to cock the weapon manually if the breech was closed. TM E9-369A explains the actual firing mechanism in detail:[5]

> *f. The cradle firing mechanism is located on the left side of the cradle. Raising the firing lever of the cradle firing mechanism forces the lug up at the end of the sear operating lever. This, in turn, pushes the sear down against the sear spring disengaging the sear lug. The firing pin holder and firing pin, thus released, are driven forward by the compressed firing spring to fire the primer in the cartridge.*
>
> *g. The firing mechanism will not operate unless the breech-block is fully closed. If the breech-block is not fully closed, the operating rod will not be fully in position against the compression of the operating spring rod. This will prevent the safety stop lever from*

TO OPERATE THE BREECH MECHANISM

a. To open.

(1) Normally, in action, the breech is opened, percussion mechanism cocked, and cartridge case extracted during counterrecoil of the gun.

(2) To open the breech manually before inserting the initial round of ammunition, grasp the breech actuating lever and squeeze the trigger to release the retaining catch. Rotate the breech actuating lever clockwise as far as it will go.

(3) Opening the breech manually may be performed either during the engaged or disengaged position of the 'SEMIAUTOMATIC-HAND' ('SEMIAUTOMATIK-HAND') catch. With the catch engaged, a strong pull to rotate the breech actuating lever is necessary. To engage the catch, pull down on the catch plunger and raise the catch in front to engage the breech actuating mechanism. The catch is disengaged when it is pressed down in front.

b. To close.

(1) Normally, in action, the breech is closed by the cartridge base tripping the extractors, thereby releasing the breech-block, which closes due to the force of the spring in the breech actuating mechanism.

(2) After the firing period, it is necessary to close the breech. This is accomplished by rotating the extractor actuating lever in a clockwise direction or operating the loading tray interlock without a round in the tray.[4]

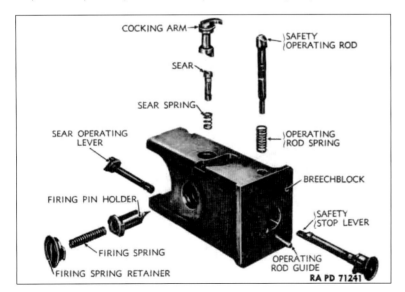

BELOW An exploded view of the 8.8cm Flak breech-block mechanism, which included the firing pin and sear mechanisms. *(US War Department)*

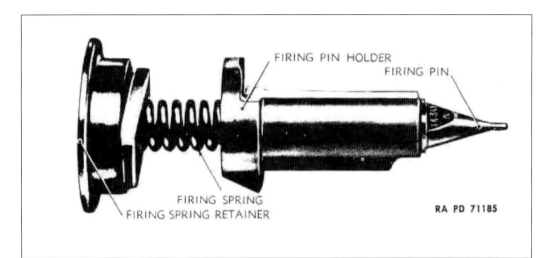

LEFT The Flak 18/36/37 were percussion-fired guns, the firing pin mechanism (seen here disassembled) striking physically against the primer in the cartridge base. *(US War Department)*

rotating, and hence will not permit clearance for the firing pin holder to move forward, thus rendering the firing mechanism inoperative.

Although somewhat complicated in narrative description, the firing mechanism of the Flak 88 was solid and reliable, with critical standards of safety. The point made in the manual about the firing mechanism not working unless the breech-block was fully closed was critical; the detonation of a shell in a partially open breech would have led to the instant destruction of the gun, and the death or injury of many of those around it.

The extractor mechanism of the Flak gun consisted of a double claw system, the claws gripping the bottom of the spent cartridge base to draw it out from the chamber (all being well). Both claws were mounted on an actuating shaft that ran out of the right side of the breech, terminating in a handle that could be rotated by the gunners.

Recoil mechanism

The recoil system for the Flak 18/36/37 featured two main parts – the recuperator cylinder and the recoil cylinder. The former was responsible for returning the gun to battery following the recoil phase, while the latter

BELOW A diagram of the loading tray interlock mechanism. Note the position of the firing lever, which was lifted upwards to fire the gun. *(US War Department)*

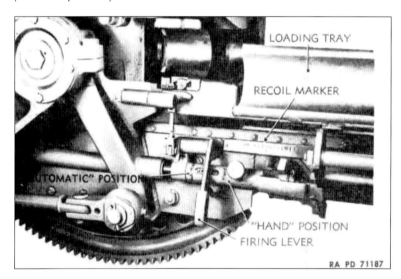

RIGHT A close-up view of the breech operating handle. Squeezing the trigger on the handle released the retaining catch. *(Author/Axis Track Services)*

ABOVE A top view of the breech, looking forward to the rear of the recuperator. The handle in the centre of the breech is the cocking lever. *(Author/Axis Track Services)*

served to slow down the recoiling barrel in the first place. In reality, the functions of each of these components were heavily interrelated.

The recuperator cylinder was the visible cylindrical structure sat atop the barrel. Inside were two cylinders: a liquid cylinder filled with thick glycerine water and a gas cylinder filled with nitrogen. The liquid cylinder sat within the gas cylinder, both running parallel with one another. The total volume capacity of the recuperator cylinder was 4.5 gallons.

When the gun was fired and recoiled, a recuperator cylinder rod and piston were driven backwards, in the process forcing, at pressure, the liquid from the liquid cylinder into the gas cylinder. The incompressible liquid squeezed the compressible gas inside the cylinder, partly providing a cushioned

LEFT Here we see the rear end of the breech-block in the side of the breech mechanism. To ensure safe firing, the breech-block had to be perfectly flush with the side of the breech housing. *(Author/Muckleburgh Collection)*

ABOVE The recoil marker scale was used to keep track of the length of the barrel recoil; excessive readings could indicate a problem with the recuperator or recoil mechanisms. *(Author/Axis Track Services)*

deceleration of the recoil energy (the recoil cylinder delivers the rest of the recoil resistance). During the gun's return from the recoil, the gas expands again and forces the liquid back into the liquid cylinder, assisting in driving the gun back to battery. The US Army manual on the gun adds a couple of interesting technical clarifications:[6]

> After several rounds have been fired, the gas and liquid have emulsified. This condition, however, does not alter the volume pressure relationship, and the liquid is still effective for its original purpose of supplying an adequate pressure seal. The ports in the end of the liquid cylinder are not throttling orifices, and the state of emulsification has no effect on the recoil action.
>
> (3) The piston rod is hollow to eliminate the vacuum which would be caused by the sealed cylinder and plug. This hollow opening also permits exit of the atmospheric air in back of the piston head. The washers are of U-shaped leather and use U-shaped brass spacers. The whole is secured by a large lock nut.

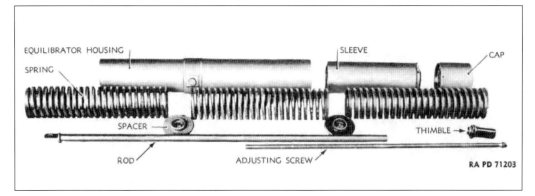

LEFT An internal view of one of the equilibrators, which would be set within a telescopic housing and worked to balance out the weight of the barrel. *(US War Department)*

ABOVE A close-up view of the recoil cylinder housing beneath the barrel. The length of recoil was self-adjusting according to the elevation of the barrel. *(Author/Muckleburgh Collection)*

BELOW The top left side of the upper carriage, including the gun bell positioned at the top of the mount arm. *(Author/Muckleburgh Collection)*

BELOW One of the 8.8cm Flak outriggers, showing the levelling jack and the stakes used to provide extra resistance against the carriage shifting under recoil. *(US War Department)*

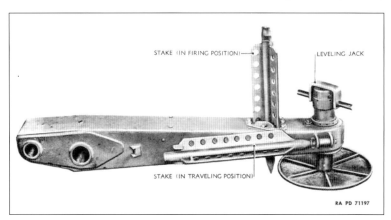

The recoil cylinder formed the other critical component for handling the gun's recoil phase. It sat beneath the barrel, filled with 9.5 litres (2.5 US gallons) of fluid at atmospheric pressure. When the gun was recoiling, the cylinder – which was attached to the gun's cradle – and its control rod were stationary, while the piston rod and counter-recoil rod moved with the breech ring. As the gun went back, the fluid in the recoil cylinder was forced through apertures in the piston-rod head, through control grooves in the recoil control rod, and through a valve in the control bushing. By squeezing the fluid through the resistant holes and grooves, against the force of recoil, the recoil force was essentially absorbed as the gun came to the end of its travel. The release of tension in the system at this stage, with the fluid returning to its original state, helped the recuperator cylinder bring everything back to its original configuration.

It must be said that the components dedicated to controlling the Flak 8.8cm's recoil forces had their work cut out. The total weight of the gun's recoiling parts was 1,435kg (3,159lb), driven backwards – in obedience to Newton's Third Law of Motion – with the same force as the outward impetus of the shell. Furthermore, these parts might have to sustain the impact of the recoil at up to 20 rounds per minute, for many minutes at a time. Note that the recoil travel distances of the gun parts varied according to the angle of elevation of the gun. At 0 degrees of elevation, the recoil was at its maximum of 105cm (41.5in), while at maximum elevation the recoil length was considerably shortened, to just 70cm (27.75in).

Mount

The mount was essentially the part of the gun that delivered its battlefield functionality. For a start, the mount gave the requisite stability needed to fire what was an extremely powerful weapon – we should remember that the Flak 8.8cm fell into the category of 'heavy' AA guns. It also carried the elevation and traverse mechanisms, making the gun trainable to the intended point of aim, and the systems for integrating the gun and its crew with the fire-control centre. Furthermore, the mount also

LEFT Side view of the Flak 37, showing the outrigger in its fully raised position. Although the levelling jack foot was large, the outrigger still required staking in place to avoid slippage. *(Author/Muckleburgh Collection)*

BELOW The locking mechanism of one of the *Sonderanhänger 202* bogies; the locking handle can be seen running up the left side of the image. *(Author/Muckleburgh Collection)*

provided the necessary mobility to move around the battlefield and do its work.

Lower carriage

At the base level of the mount were the two two-wheeled bogies – in effect a front and rear chassis – used to make the gun mobile. Typically the bogies were fitted with pneumatic tyres, although solid rubber versions are also seen, and the wheels of the front chassis were steerable; the steering could be manually disabled if required. All the wheels had leaf suspension and air brakes, the latter actuated by the half-track driver pressing his foot on the brake pedal; the rear bogie was also fitted with a crude steel seat from which a member of the

LEFT When folded up into the travelling position, the outriggers were secured with a locking pin and a length of chain. *(Author/Axis Track Services)*

gun team could operate a handbrake if required. To deploy the gun from the limber, jacks beneath the gun's lower carriage were first deployed to relieve the pressure of the gun, then the bogies could be disconnected and withdrawn. To assist, a special chain winch with lifting chain was mounted above the axles. The interface between the gun and the bogies was the bottom carriage, which when connected to the bogies acted as a form of chassis for the gun.

The bottom carriage itself was a welded and riveted box-type construction, with folding stabilising outriggers on each corner. These hinged outriggers were integral to the bottom carriage, and during travel were folded into an upright position. When the gun was placed in the firing position, the outriggers were released

BELOW The mudguards of the *Sonderanhänger 202* featured storage fittings plus the mounts for the data cable drums. *(Author/Muckleburgh Collection)*

BELOW LEFT The front bogie of the *Sonderanhänger 201*. The bogie chain was used to take the weight of the gun off the limber locking jaws before dismounting. *(US War Department)*

and folded down to create a very stable cruciform platform for the upper carriage and gun; the outriggers were secured with the half-round locking pins plus metal stakes driven into the ground through the strut. At the end of each outrigger was a screw-type levelling jack, so that the gun could be set level even on moderately uneven ground. Other components of the bottom carriage were the large gun pedestal, bolted to the base in the centre, plus handwheels for levelling the top carriage – via linkages to the top carriage, the handwheels could 'tip the top carriage about the two centers of rotation, thereby aligning the gun trunnions at a horizontal position. A level indicator is provided on the pedestal.'[7] At the rear end of the bottom carriage was also the data transmission junction box, which connected the gun to the centralised fire-control network.

Upper carriage

The upper carriage of the Flak 8.8cm held the gun itself, plus most of the mechanisms and additional apparatus for operating and firing the gun. The following parts constituted the upper carriage:

- The cradle for mounting the gun, with the recuperator cylinder, recoil cyclinder, barrel, rammer and loading tray
- Two spring equilibrators, used to balance out the preponderance of the barrel weighting towards the muzzle
- The traverse mechanism
- The elevation mechanism
- The receiver dials for traverse (azimuth) and elevation, the dials indicating how the traverse and elevation should be applied to bear on to the target
- The aiming assembly
- The fuze-setting machine, for setting the time-detonated fuzes of the shells to the correct interval
- The gunner's seat
- A box holding the flak-aiming telescope.

The gun cradle was a rectangular trough made from welded and riveted sheet steel, the trunnions on the sides slotting into the arms of the cradle either side of the gun. Two spring equilibrators were fitted to the underside of the

ABOVE A close-up of a carriage-leveller handwheel. The two handwheels tip the carriage around two centres of rotation. *(Author/Muckleburgh Collection)*

cradle, visible in two thick telescopic housings. Note that other fittings directly to the cradle are the breech operating cam and the auxiliary trigger mechanism, both fixed to the rear right

BELOW This fascinating diagram shows the 88's carriage-levelling mechanism; it was essential that the gun trunnions were horizontal before the gun was fired. *(US War Department)*

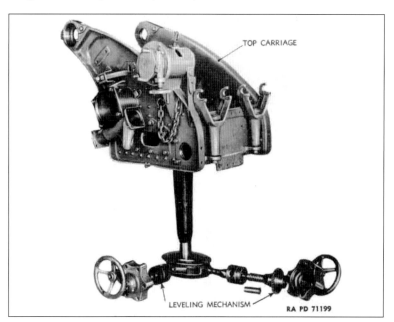

LEFT The disconnected muzzle rest of the Flak 18/37, showing the barrel locking arm in its open position. *(Author/Muckleburgh Collection)*

side, while on the opposite side we find the loading tray, firing lever and recoil marker.

For bringing the gun on to target, the upper carriage featured two large handwheels on the right side of the mount, one for elevation and one for traverse. Both mechanisms could be operated in either high- or low-speed settings, depending on the tactical situation, and a gear selector lever was provided by the side of each wheel for switching between the two modes.

As the gun was capable of making two 360-degree rotations in the traverse – any more than that and the data transmission cable would get tangled up and damaged – an indicator set above the traverse handwheel told the gunner where he was in the rotation, and a spring-stop mechanism at the left side of the top carriage prevented the gun from moving beyond the permitted turn. With the elevation mechanism, 'Motion is transmitted through the handwheel, through gears, to the elevating pinion which engages the elevating rack, thereby elevating or depressing the gun.'[8] Note also that the upper mount was fitted with a clutch to disengage the elevating mechanism from the elevating rack (a curved, toothed fitting just beneath the cradle) when the gun was being transported; leaving the elevating mechanism engaged would result in the shocks of road travel being transmitted directly to the teeth of the elevating rack. One

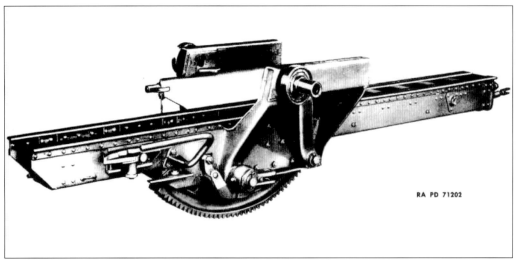

RIGHT The 8.8cm Flak gun cradle. Fitted to the rear right side of the cradle were the auxiliary trigger mechanism and the breech operating cam. *(US War Department)*

FAR LEFT The pedestal provided the structural connection between the cruciform platform and the carriage. The two handwheels are the controls for the gun-levelling mechanism. *(Author/Axis Track Services)*

LEFT One of the 8.8cm Flak's stabilising outriggers in the upraised position, with the fixing stakes secured at the sides. *(Author/Axis Track Services)*

other purpose of the clutch was to ensure that the teeth of the elevating mechanism were properly aligned before they were engaged, to prevent wear and tear on this component that was so critical to the gun's accuracy.

As mentioned in Chapter 1, the azimuth and elevation receivers on the Flak 18/36/37 changed over time according to the gun type. The Flak 18 and 36 utilised the *Übertragungsgerät 30* dials to connect the instructions from the fire director to the traverse and elevation of the gun itself. How this worked, both mechanically and practically, was explained in the US War Department Technical Manual:[9]

(1) The azimuth and elevation indicators on the gun carriage are identical. Each indicator consists of a cylindrical case containing three concentric rings of 10-volt electric lamp sockets. Each socket has an individual positive electrical connection. All sockets in each indicator have a common negative connection.

(2) Three pointers are pivoted at the center of each indicator, one for each circle of lights. At the end of each pointer is a translucent index. The two inner indexes are each wide enough to cover one light. The outer pointer can span two lights. The

BELOW LEFT The traversing handwheel, turned clockwise for right traverse and anti-clockwise for left traverse. *(Author/Muckleburgh Collection)*

BELOW The traverse mechanism housing, located on the base of the cradle next to the fuze-setting mechanism. *(Author/Muckleburgh Collection)*

DESIGN AND BASIC OPERATION

LEFT The toothed elevating arc, set on the underside of the breech. *(Author/Muckleburgh Collection)*

CENTRE The elevation clutch disassembled; this mechanism disconnected the elevating mechanism from the elevating arc for transit. *(US War Department)*

pointers are geared together at a ratio of 1:10:100; the shortest pointer moving 1 turn to 100 of the target pointer. The pointers are mechanically coupled to the azimuth and elevation drives of the gun carriage. A knob at the center of each indicator is used for synchronizing the indicator arms with the gun prior to operation.

(3) The case is bolted to the top carriage and has a cover which consists of a 10-armed spider which supports a translucent plastic sheet cupped within the spider. A dowel on the case fits a hole in the rim of the spider, permitting assembly in only one position. This assures maximum visibility of the lights. Two spring clips clamp the cover in place. At the center of the spider is a chained cap which protects the adjusting knob of the indicator.

b. Operation.

(1) Sight the range finder of the stereoscopic director on a distant aiming point (a distant terrestrial object or a celestial body) and bore sight the guns of the battery on this target. If the lighted lamps of the azimuth and elevation indicators of each gun cover the lighted lamps the gun is oriented with the director. If adjustment is necessary, remove the metal cap and engage the knob with the cross piece in the shaft, turn until the dial is blacked out, and release the knob.

(2) The gun is thereafter operated as the lights flash on around the circle by turning

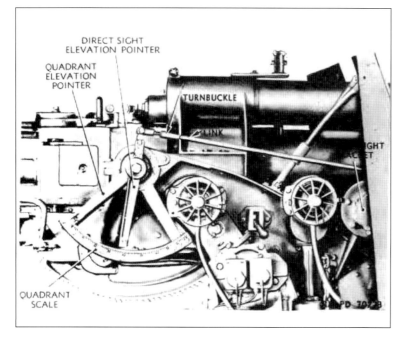

LEFT The elevation quadrant was the other component of the 8.8cm Flak's direct-fire capability. It was graduated in 0.25-degree intervals from –3 degrees to +85 degrees. *(US War Department)*

the elevating and traversing handwheels to keep the lamps blacked out.

Although the *Übertragungsgerät 30* served its purpose, it was a difficult system both to use and maintain. Blown bulbs in the concentric rings of lights could confuse the direction, and bright direct sunlight could also make seeing the lights difficult in the first place. Thus, in the Flak 37, the system was replaced with the *Übertragungsgerät 37*, in which the information from the predictor automatically moved pointers on the azimuth and elevation receivers, and in turn the gunlayer simply had to rotate the elevation and traverse wheels so that he matched a second set of pointers with the pointer direction set centrally. It was a simpler, and therefore faster, system of operation.

The rammer system on the Flak 8.8cm was important for loading the gun when it was at a high angle of elevation, which presented the breech at an awkward angle for the loader. The rammer mechanism was located on the left rear side of the barrel, mounted to the cradle. It operated in a similar way to the recuperator cylinder, through hydropneumatic principles utilising the interaction of gas and liquid. The fluid/gas was placed under pressure by the mechanical action of the gun returning to battery under the force of counter-recoil, at the same time – through a rack and pinion linkage – pushing out a ramming arm in the opposite direction, essentially cocking this mechanism for loading. To use the ramming function required an interaction between the hand-operated loading tray – mounted on two supporting lugs on the left side of the cradle – and the rammer head, which was rotatable at the end of the rammer arm. As a rough description, the loader placed the shell on the loading tray, then swung the loading tray to align the shell with the breech, an action that released the loading tray interlock and permitted the rammer mechanism release, using the force of the expanding gas to pull the rammer arm forward and seat the shell.

Another important mechanism on the Flak 8.8cm guns was the fuze setter, mounted on the left side of the carriage. This device was used to set the time-activated fuzes on shells quickly in response to data being fed

LOADING THE FLAK 8.8CM

27. TO LOAD.

a. Place the shell on the loading tray and swing the tray in line with the axis of the bore of the gun. At this point the loading tray interlock is released and the expanding gas forces the rammer cylinder back along the piston; thus the rammer arm is rapidly withdrawn, seating the round. Swing the empty loading tray back to its original outboard or loading position.

l). When firing at angles above 45 degrees, set the buffer valve to 'QUICK' ('SCHNELL') by turning the valve clockwise. At angles below 45 degrees, the valve is set to 'NORMAL' ('NORMAL') by rotating the valve counterclockwise.

28. TO FIRE.

a. With the loading tray interlock set at 'AUTOMATIC' ('AUTOMATIK'), the gun will fire as soon as the loading tray clears the path of recoil and is returned by hand to its outboard or firing position.

b. With the loading tray interlock set at 'HAND' ('HAND'), the gun must be fired by performing either one of the following steps:
   (1) Raising the firing lever on the left side of the cradle.
   (2) Pulling the auxiliary trigger on the right side of the cradle.

29. TO RECOCK.

a. In case of a misfire, it will be necessary to recock the percussion mechanism by rotating the cocking lever in a counterclockwise direction as far as the word 'WEIDERSPANNEN', which means 'RE-COCK'. Then return the cocking lever to its original position at 'FIRE' ('FEUER'). Fire the gun as described in paragraph 28 again; and if the gun again misfires, wait 10 minutes and then unload as described in paragraph 30.

30. TO UNLOAD.

a. Open the breech. If the extractor does not eject the shell, grasp the shoulder on the cartridge base and withdraw it from the breech recess. Then reload the gun.[10]

LEFT The breech end of the Flak 37, showing the silver-coloured breech-block slid over to the left in its locked position, ready for firing. *(Author/Axis Track Services)*

ABOVE A shielded Flak 36 in its travelling position. The gun could be fired from its wheels, but its accuracy in this configuration was compromised. (US War Department)

either electrically from the stereoscopic director or telephoned from the auxiliary director (see below for further details about sighting and fire control). The fuze setter had two setting ports into which two shells at a time could be seated, nose down, connecting the fuzes at the tip of the shell with an adjusting mechanism to alter the fuze timing. It consisted of:

> . . . a plate containing three concentric rings of 10-volt electric lamp sockets. Each socket has an individual positive electrical connection. All the sockets have a common negative connection. Three pointers are pivoted at the center of the circles, one for each circle of lights. At the end of each pointer is a translucent index. The two inner indexes are each wide enough to cover one light. The outer pointer can span two lights. The pointers are geared together at a ratio of 1:10 :100, the shortest pointer moving 1 turn to 100 turns of the longest pointer. The pointers are geared to the fuze dial and to the setting ring of the fuze setter.
>
> 3) For use with data telephoned from the auxiliary director, the fuze setter has a scale graduated from 15 to 350 degrees. The safe position is marked with a cross. [. . .]
>
> 4) The setting crank at the front of the fuze setter turns the pointers and the fuze dial. The crank at the side of the fuze setter turns the inertia flywheel which stores up mechanical energy for cutting the fuzes. The release lever releases the round after the fuze is cut. The cable receptacle from the fuze setter extends to the terminal box at the front of the top carriage.
>
> Operation.
> (1) Turn the setting handwheel to black out the lights (in operation with the stereoscopic director) or to aline [sic] the fuze scale in accordance with the values announced by telephone (in operation with the auxiliary director). Turn the power crank, thus storing up energy in the flywheel, and keep the crank turning at a uniform rate.
>
> (2) Thrust the round sharply into one of the cups of the fuze setter, thereby engaging a toothed clutch which rotates the adjusting pin. This makes two complete turns before being automatically disengaged. The round is held in position by a key which rides in a circular groove at the bottom of the fuze. This key is tripped by a lever at the top of the setter and the round is released. Two fuzes may be cut at one time.[11]

Set just above the fuze setter on the carriage was an electrically operated bell. This performed several alert functions. For example, it could be used to indicate that the gun was now on target and that the shot should be taken. Or it might be applied to indicate time intervals, to establish a regular and repeated pattern of firing.

The operation of azimuth and elevation receivers, and of the fuze setter, relied upon a smooth flow of data from the fire-control systems, about which we shall say more shortly. The main electrical point of contact was located on the end of the rear trail section. Here, beneath a metal cap, was an electrical connector with 104 pins, into which the cables and plugs from the azimuth and elevation receivers and the fuze-setting mechanism were connected.

Sighting and fire control

The power of the 8.8cm Flak gun was nothing if the shells were not fired to the right place and at the right moment in time. The subject of fire control is a complex one, involving not only the sighting and fire-control systems of individual guns, but also the systems used to coordinate multiple guns and batteries and, in the case of AA fire, how these subsystems connected to the larger network of regional and even national fire control. Here, however, we can look into the details of the core sighting options for direct fire and the typical fire-control mechanisms used for indirect fire, either for AT or for AA roles.

LEFT The 104-pin data transmission receptacle located at the end of the rear trail, for connection from the fire-direction equipment to the gun receivers. *(US War Department)*

It is worth reminding ourselves of some of the complexities of artillery fire control. The simplest relationship between a gun and its target is if the target is stationary, well within the gun's effective range, and the gunlayer is able to acquire that target visually through an optical sight. In this case, the gunlayer usually has little more to do than to place the target at the right point on a sight reticle – allowing a little for the fall of the shot or perhaps for windage – then

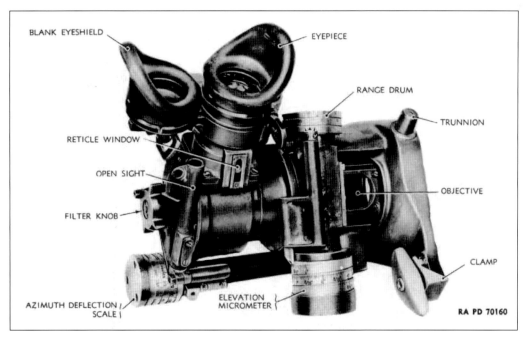

LEFT The ZF.20E telescopic sight, which was mounted on the right side of the gun. It was used, along with the elevation quadrant, specifically for direct fire. *(US War Department)*

ABOVE A close-up view of the toothed elevating arc beneath the gun carriage. *(Author/Muckleburgh Collection)*

engage in direct-fire gunnery. Unfortunately for most artillery crews, life is rarely that simple. As soon as the target begins to move, then *lead* will need to be applied to the aiming solution; i.e. the gunlayer will need to fire at the point where the target *will be* by the time the shell has travelled from the muzzle to the impact point. This equation involves finding a trigonomic firing solution that accounts for factors such as: (1) the speed of the shell in flight; (2) the time it will take the shell to cross the required space; (3) the drop and drift of the shell; (4) the target's direction of travel/changes in altitude; (5) the target's speed of travel; and (6) estimated changes in the target's direction of travel (e.g. if the target is making evasive manoeuvres. One of the advantages of the 8.8cm Flak's high muzzle velocity was that it reduced the need to apply lead to a target, yet time still came into play. The flight time of a shell against an enemy T-34 2,500m (2,734yd) away would be in the region of 3–4 seconds at a muzzle velocity of 750m/sec (remembering that the velocity is dropping from the moment the shell leaves the muzzle of the gun). When fired at an aircraft at 7,620m (25,000ft), by contrast, the time of flight to target could be in the region of 25 seconds. Fire control therefore involves, to a high degree, technological *prediction*, firing where the target is predicted to be, not where it actually is. It also requires an accurate estimation of range, not something human beings are particularly good at with the naked eye once an object is more than about a kilometre away.

Now also remember that in many cases – particularly involving AA fire at high-altitude targets, during night-time hours or when delivering indirect ground fire – the gun crew themselves could not actually see the target. In these instances, technologies external to the gun crew provided the information that enabled them to lay and fire the gun. These technologies included radar, advanced rangefinders and director devices. Furthermore, the information flowing into the gun receiver systems from the external fire control is rarely for a single gun; often centralised information must flow out to several guns or several batteries to coordinate the fire. Add to this increasingly complex business the exigencies

RIGHT Taken from beneath the front bogie of the *Sonderanhänger 202*, this photo shows the foot of the levelling jack and, either side, the suspension leaf springs. *(Author/Muckleburgh Collection)*

RIGHT **A ZF.20 optical sight in situ in its storage box beneath the barrel.** *(Author/Axis Track Services)*

of quick fire against a responsive enemy, and the human pressures of coping with being under fire, then we can appreciate that the Flak 8.8 guns needed straightforward sighting and fire-control systems to make meaning out of their ballistic performance.

Direct-fire sighting

For direct fire, either against ground targets or against visible aerial targets that could be acquired visually, the primary instrument for fire control would be the Flak ZF.20 or ZF.20E, the 'ZF' part of the name referring to *Zielfernrohr* (telescopic sight). The difference between the two devices is that the ZF.20E featured a range drum, whereas the former did not – the implications of this will be presented below, but the core of our description here will focus on the ZF.20E.

The optical sight was stored in the box located just beneath the barrel. When direct fire was required, the sight would be mounted on a bracket on the right side of the gun, directly in front of the gunner's seat. The primary target viewing component of the ZF.20E was the 4× power elbow telescope, offering a 17 degrees 30 minutes field of view and a partial crosshairs reticle with an inverted 'V' aiming point in the middle. The reticle also included some guidance on lead calculations in the display, and the display could be illuminated via a lamp attachment. Sighting in different daylight and climate conditions was aided by the incorporation of four filters – clear, green, light neutral, dark neutral – into the viewfinder; the gunner could rotate through these filters using the filter knob on the side of the sight. A blank eyepiece was provided by the side of the telescopic eyepiece, purely as a place for the operator to rest his eye – a nod to ergonomic design in an age not known for its respect of human comfort.

RIGHT **The 4× power optical lens of the ZF.20. The dial to the left is the elevation knob.** *(Author/Axis Track Services)*

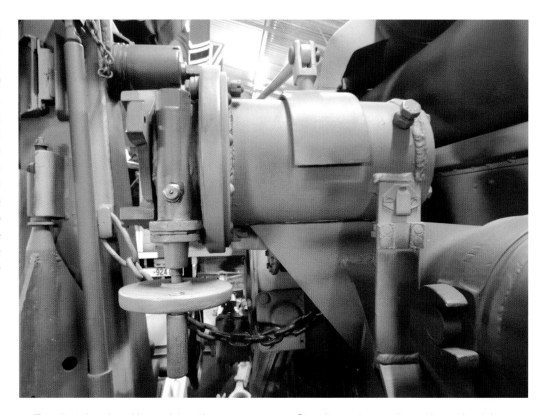

RIGHT The aiming assembly mechanism was used when there was a loss of command control, and when fighting against ground targets. Turning the handwheel (bottom left) transferred elevation readings to the elevation degree scale next to the breech. *(Author/Axis Track Services)*

BELOW The elevation receiver of the *Übertragungsgerät 37* system. The gunner simply had to keep the pointers aligned with the directional dials, which were automatically adjusted via the fire director. *(Author/Muckleburgh Collection)*

To adjust the aim with precision, the ZF.20E featured a deflection mechanism, with the deflection scale adjusted by rotation of a deflection knob from 250 mils left to 250 mils right. An angle of site mechanism also then rotated the telescope and deflection mechanism within the main sight housing, the scale for this feature graduated in 100-mil intervals from plus to minus 200 mils. The range quadrant and elevation mechanism in turn rotated the entire telescope and angle of site mechanism in the housing.

Standing out conspicuously on the right side of the gun is an elevation quadrant, looking rather like part of a sextant, featuring a curved metal scale and a scale pointer. The US War Department Manual usefully unpacks its function:[12]

(1) The elevation quadrant is used in direct fire sight (ZF.20E or ZF.20) for laying the gun in elevation. The elevation quadrant includes a quadrant, an outer pointer, an inner pointer, and a link to the telescopic sight bracket.

(2) The quadrant is centered on the cradle trunnion and is fastened to the mount. In operation the quadrant remains stationary. The quadrant is graduated in 0.25-degree intervals from minus 3 degrees in depression to plus 85 degrees in elevation.

(3) The outer pointer (quadrant elevation pointer) is fastened to the cradle trunnion. It bears an index which alines [sic] with the quadrant scale, indicating the elevation of the gun.

(4) The inner pointer (direct sight elevation pointer) pivots on the cradle trunnion and moves with the bracket of the telescopic sight. The pointer has an index line which alines [sic] with the index of the outer pointer.

(5) The link connects the inner pointer with the bracket of the telescopic sight and has a turnbuckle for alinement of the sight with the gun bore.

(6) When the elevation quadrant, the link, and the telescopic sight are in adjustment, the gun may be laid in elevation by setting the desired elevation or range in on the telescopic sight and matching the pointers on the elevation quadrant.

Once the telescopic sight unit had been bore-sighted to the gun – meaning that the gun and the sight were perfectly aligned to the aim point – then the gunner could begin using the sight for direct-fire shooting. He sat on the gunner's seat behind the telescope with the traversing handwheel positioned to his left, and with the angle of site scale and micrometer set at zero. Then the elevation for range in degrees and $\frac{1}{16}$ degrees would set in on the elevation scale or in metres (on ZF.20E only) on the range drum. If the gunner had to build in deflection, he could do that visually by compensating against the sight reticle, but that would only work if the deflection was slight. Larger deflections would have to be set in mils on the deflection drum. The gun was laid for elevation by matching both pointers of the elevation quadrant, while the gunner tracked the target left and right with his traversing handwheel.

Fire-control mechanisms

For the 8.8cm Flak, the three core fire-control devices at the beginning of the war were the *Rundblickfernrohr 32* (Rbl. F. 32; Panoramic Telescope 32), the *Kommandohilfsgerät 35* (KDO. HI. GR. 35; Auxiliary Director 35) and the *Kommandogerät 36* (KDO. GR. 36; Stereoscopic Director 36). The Rbl. F. 32 was a 4× power fixed-focus telescope fitted in a special mount atop the gun's recuperator. Its principal application was to enable the gun to perform the indirect-fire role against ground targets, aligning the barrel of the gun with the information flowing through from the external directors. Because indirect fire was critically concerned with the degree of elevation to apply to the gun barrel (to allow the correct arc of fire on to the target), the wheels on the device were chiefly focused upon azimuth adjustment,

ABOVE A close-up of the gun bell on the upper carriage. The bell was essential to transmit instructions over the din of battle, and was used to command fire, dictate firing intervals or implement other agreed procedures. *(Author/Muckleburgh Collection)*

LEFT The storage box for the ZF.20 telescopic sight (top) plus an electrical data connection port below. *(Author/Muckleburgh Collection)*

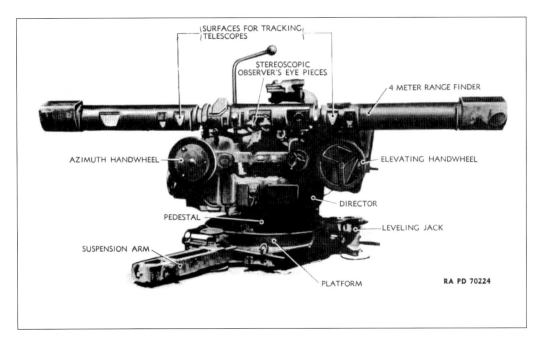

RIGHT The KDO. GR. 36 was the principal fire director for the 8.8cm Flak guns. Its rangefinder scale ran from 500m (546yd) to 50,000m (54,680yd).
(US War Department)

with the two scales on the vertical barrel of the telescope graduated in 100-mil intervals. A US Army *Catalog of Enemy Ordnance* from 1945 explained the configuration of the sight:

> When mounted in their respective sockets with the azimuth scales set to zero, the line of sight of the panoramic telescope on the predictor is 180° from that of the predictor telescope. The reason for the eyepiece of the gun sight being 90° from the axis of the gun is for convenience as the operator can stand at the right side of the gun and look into the sight at right angles to the axis of the gun.'[3]

The KDO. GR. 36 was a very different bit of kit compared to the Rbl. F. 32, not least in terms of scale and mounting. It consisted of two main pieces. First, there was the director unit that provided centralised predictor control to a gun battery, feeding the guns a continuous flow of information about the aerial target, specifically quadrant elevation, future azimuth and time of flight of projectile, expressed in fuze units. To collect and transmit this data, the director had to gather the following information about the target: (a) present angular height; (b) present azimuth; (c) present slant range.

The director in itself was a large piece of

RIGHT The elevation and traverse mechanisms of the KDO. GR. 36 fire director, which kept the reticle of the elevation and azimuth tracking telescopes on the target.
(US War Department)

kit, operated by a crew of about 11 men and transported on its own double bogie similar to the bogies used for transporting the gun; when deployed for action it sat on three levelling feet and the two suspension arms that were used to attach it to the bogies. Atop the director was fitted the *Raumbildentfernungsmesser (Höhe)* – Em. 4m. R (H) – a 4m stereoscopic rangefinder secured to the top of the director via two clamps. The rangefinder used optical principles of triangulation to obtain altitude information about the target. It had a visual magnification of 12× and 24× with a range scale reading that ran from 500m (550yd) to 50,000m (55,000yd). In addition to the stereoscopic eyepiece in the centre of the rangefinder, two tracking telescopes were mounted, one at either end, each with a crosshairs reticle pattern to enable the crew to stay on target. At the right end of the rangefinder was also a manual scale device for obtaining the approximate height of the target.

Note that the KDO. GR. 36 was largely replaced by the middle of the war by an updated version, the KDO. GR. 40, which became the main director for the 8.8cm Flak batteries during the critical years of fighting against the Allied strategic bombing campaign, 1943–45. The problem with the KDO. GR. 36 was that it struggled to feed good information to the gun crews if the target was not obligingly flying straight and level (i.e. it was taking evasive manoeuvres or adjusting its flight path for an attack) and if the target was flying at anything more than 643km/h (400mph). By the later years of the war those performance parameters were a problem, especially as Allied bomber crews adjusted their tactics in response to German AA and fighter procedures and as Allied fighters started to exceed the speed limitations of the director. The improved KDO. GR. 40, therefore, made adjustments that not only allowed the system to track targets flying in excess of 643km/h (400mph), but could also hold on to aircraft making turns, or at least gradual ones. There were also updated versions of the rangefinder devices, making improvements to target acquisition and holding and to the general practicality of the use.

The data that was provided by the KDO. GR. 36 or KDO. GR. 40 could be fed directly to

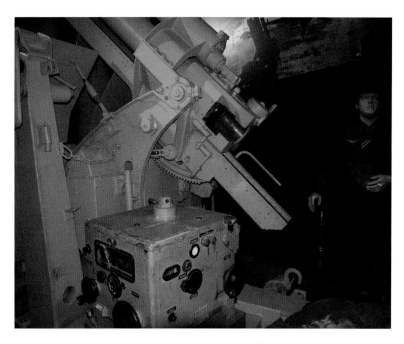

ABOVE An 8.8cm Flak gun alongside a *Kommandohilfsgerät 35* auxiliary fire director. The instrument weighed some 20kg (45lb), and was transported on a wheeled trailer. *(JustSomePics)*

the Flak guns' elevation and azimuth receivers via electrical connection, as noted above. For situations in which this connection was not possible, or where one of the large directors was not available, there was also an alternative in the form of the KDO. HI. GR. 35. This was a simpler piece of equipment used to gather data on altitude, elevation and fuze time, data that was then telephoned across to the gun crew for manual setting – in fact, the telephone system was the only electrical current required to operate the whole director. The KDO. HI. GR. 35 was transported on a two-wheel trailer, and when deployed was lifted off the trailer and set up on a three-leg collapsible stand fitted with three levelling screws in the mounting head. The KDO. HI. GR. 35 calculated the firing solution using the angular rate method, explained by the US War Department Manual as follows:

The slant range prediction is approximated by adding range rate times future time of flight to the present slant range which is obtained from a 4m stereoscopic range finder set up nearby. The super elevation and fuse are taken from three-dimensional cams, positioned by future angular height and

RIGHT The front bogie of the *Sonderanhänger 202*, with the handbrake on the left side, next to the winch for raising the gun off the bogie locking hooks. *(Author/Muckleburgh Collection)*

BELOW This view of the side of the 8.8cm Flak shows the Receiver C fuze-setting receiver dials, just next to the equilibrator. *(Author/Muckleburgh Collection)*

future slant range. Lateral, vertical, and range rates are measured by tachometers and are manually matched. Deflections are computed by multiplying present angular velocity by present time of flight.[14]

Portable fire-control kit

A battery of 8.8cm guns was at the centre of a range of skills, work and technologies. In addition to the larger fire-control devices outlined above, there were also a variety of smaller man-portable pieces of kit often seen around the guns in action. For example, in contrast to the sizeable Em. 4m. R (H), there was the man-portable 0.7m Model 34 optical coincidence rangefinder, which provided range values via two objective lenses feeding to a common eyepiece. This item was carried in a case by one man, and when set up it utilised a metal brace frame so that the user could stabilise the rangefinder against his shoulders, giving him a better platform for observation. To measure the range of a target, the operator would simply align the sight on to the target, and centre the target in the optical viewer. The image would be shown as 'split' (by virtue of seeing the same object through two objective lenses), and the operator would then turn the range knob until the images focused into one; at this stage the operator could then read the range values in the field of view and pass the information over to the gun team.

Another device, seen mounted on either tripod legs or a simple 'trench mount' spike monopod, was the aiming circle, used for measuring angle of site, for declinating and determining azimuth angles, and for spotting. The same tripod could also be used for mounting the 10× power battery commander's telescope, a scissor-type dual eyepiece telescope that proved a useful tool for observation and also the facility for measuring angles of site and azimuths. (Note that this

RIGHT **The Model 34 was a simple one-man stereoscopic rangefinder, particularly practical for rapid range calculations against fleeting or fast-approaching targets.** *(US War Department)*

device, and the rangefinder, are seen in German service in all manner of contexts, not just in the support of 8.8cm batteries.)

The technology and scientific understanding that went into putting an 8.8cm shell accurately on target at ranges extending over thousands of metres was, and remains, intellectually inspiring. What is equally impressive is that the gun crew and those operating the directors and rangefinders had to perform precision tasks within the chaos of battlefield conditions, and under the exigencies of life-or-death confrontations.

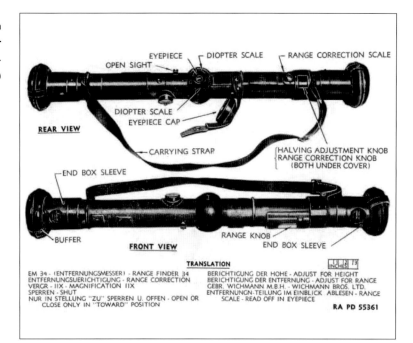

RIGHT **The battery commander's telescope was a popular piece of observation kit for artillery commanders during the Second World War. The two telescope arms were hinged, so they could be positioned horizontally to increase the stereoscopic effect.** *(US War Department)*

BELOW **A close-up of the radio headset port, positioned for use by the gunner operating the traverse mechanism.** *(Author/Muckleburgh Collection)*

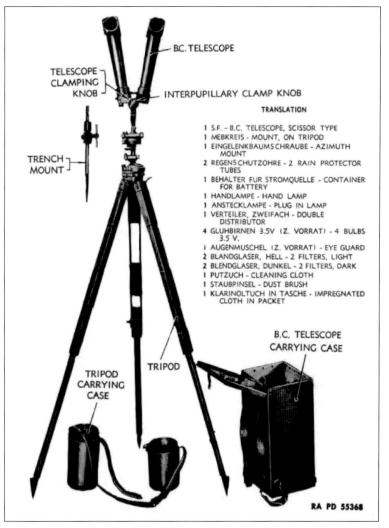

DESIGN AND BASIC OPERATION

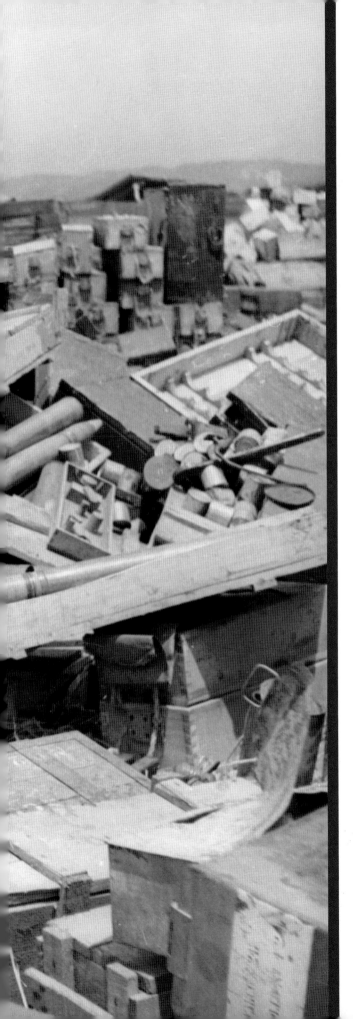

Chapter Three

Ammunition

The ammunition used by the 8.8cm Flak guns broke down into two basic categories – high-explosive (HE) and armour-piercing (AP). Within that broad division, shell types were divided by warhead type, explosive fillings and fuze configurations, to give the intended effect on target.

OPPOSITE A British soldier roots carefully through an abandoned stockpile of German ammunition – what appears to be a mix of 8.8cm and 105cm shells – in Tunisia in 1943. *(AirSeaLand Photos/Cody Images)*

ABOVE A young *Flakhelfer* prepares to load a HE shell at an 8.8cm battery in Berlin-Karow in January 1944. Loading the shells continually required great upper body strength and endurance. *(Heinz Radtke/Family Archive Norbert Radtke)*

RIGHT A German gunner loads an 8.8cm Sprgr Patr L/4.5 high-explosive shell to deliver fire against Allied invasion forces in France in 1944. *(Bundesarchiv, Bild 101I-496-3491-36/ Röder/CC-BY-SA 3.0)*

The 8.8cm Flak guns fired unitary ammunition, simply meaning that the shell case, propellant and warhead came as one fixed unit, fuzed and ready for firing, as opposed to some heavier artillery types that were loaded with separate propellant and warhead. The shells usually arrived with the front-line troops packaged either in special three-round crates, principally made of wicker (a good non-conductive and non-abrasive material), but with a metal cap and securing bands, or in single-shell steel cases.

Firing a unitary cartridge was integral to the Flak gun's performance. Such rounds facilitated fast, continual reloading – requisite for responding to an oncoming surge of Allied armour or for putting a mass barrage up into an American or British bomber force – and they were more convenient to store and transport. The gun could also be brought more quickly into action from a standing start without the need to make prolonged special preparations of ammunition.

As noted in Chapter 1, the shells for the 8.8cm Flak were basically as heavy as could be permitted for a unitary shell and for practical and sustainable one-man loading. Yet the 8.8cm Flak ammunition made no compromises when it came to power or destructive capability. Although the Flak gun itself was responsible for putting the shell on to target, it was the warhead that had to do the ultimate work, and it did this work devastatingly well. The respect that the Allied tank crews had for the 8.8cm Flak came from witnessing, first hand, the way its shells could punch through and dispatch even the heaviest armoured vehicles, or from flying through the fragment-spattering turbulence generated by hails of 8.8cm shells detonating at high altitude.

Shell cases

The overall length of the shells for the Flak 18/36/37 was 863–962mm (33.9–37.8in), although the bulk of this visible length was accounted for by the cartridge case, which always measured 568mm (22.4in). The cartridge case was not simply a practical container for the propellant, but was central to the gun's working efficiency. It had to be strong enough not to rupture under the violent pressures of propellant detonation, but also flexible enough to provide obturation – the case swelling to fit the chamber as the gun was fired, and thereby forming a gas-tight seal for the chamber. The cartridge cases had to be resistant to rust and to the invasion of moisture that would dampen the propellant and cause a misfire; a watertight seal was particularly important around the primer in the base of the cartridge case and around the warhead itself. Similarly, the case also had to keep out dust or other foreign bodies. Its obvious central role was to carry the propellant as safely as possible – a sound case design should dramatically limit the possibilities of a shell detonation anywhere other than in the gun itself.

The base of the shell case held the primer, either a C/12 nA St percussion type (meaning it was detonated manually by the impact of a firing pin) for the Flak 18/36/37/41, or a C/22 electrically fired type in the AT and tank versions. The percussion primers did not present as a visible cap in the case base, as with most Allied shell primers, but instead were formed with a steel body that was thinned in the section struck by the firing pin – the firing cap was located just behind this section. When the gun was fired, the firing pin struck the base of the primer, denting the cap and triggering the percussion cap inside, which in turn detonated the contents of the primer unit that, following the sequence, sent a hot flash of flame into the main propellant, launching the shell. The electrically fired primers worked in one of two ways. In the standard version, the primer held a contact stud in its base, this being connected internally to a filament wire that ran through the explosive primer compound. When the shell was loaded into a

LEFT The shell here, recovered from a Belgian battlefield, is the 8.8cm KwK 43 L/71, a powerful armour-piercing munition that delivered a muzzle velocity of 1,000m/sec (3,280ft/sec). *(Paul Hermans)*

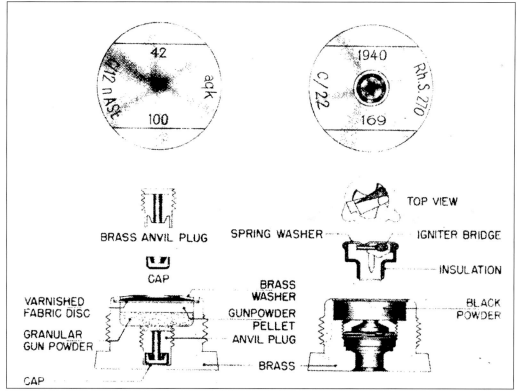

LEFT A diagram from an Allied ammunition manual shows a C/12 nA percussion primer (left) and a C/22 electric primer side by side. *(US Army/Air Force)*

RIGHT The bottom of this 8.8cm shell case displays the design number (6347), the manufacturer's stamp (ck) and the year of manufacture, plus a note on the calibre and intended weapon. *(Author/Axis Track Services)*

gun, an electrical contact pin in the gun's firing mechanism was placed into contact with the stud. On firing, an electrical contact was sent through the pin, through the stud and into the filament wire, which instantly reached white heat and detonated the priming compound. Another, more reliable, type of electric primer ignition involved linking the filament with an induction coil in the base of the primer, rather than a contact stud. On loading, this came into contact with a corresponding coil in the gun's breech-block, which provided the electrical charge that passed through the primer induction coil and heated the filament. The chief advantage of this system was reliability, as wear and tear and dirt intrusion in the pin-and-stud arrangement could cause misfires. On the other hand, as Ian Hogg explains, with induction-fired cartridges: '[These] operated successfully with a gap of as much as 1mm (0.04in) between the breech face and the cartridge case and primer.'[15]

The 8.8cm shell case was of a rimmed design, the rim providing the extractors with a grip edge for drawing out the spent shell case. The actual material used in the construction of the case varied over time, dictated largely by issues within German industry that meant the optimal material – cartridge brass (70 per cent copper and 30 per cent zinc) – had to be modified or changed entirely to meet the realities of raw material supply. (The brass shell cases can be identified by the number 6347 stamped into the baseplate.) The shell cases for the 8.8cm Flak were indeed originally pure cartridge brass, but in the second half of the 1930s, as military production levels were ramped up in expectation of war, cheaper and more accessible materials were tried. In around 1937, therefore, a drawn steel version was introduced. Steel was cheap and available, but it was also prone to rust and, if not made well, splitting, so various forms of plating were applied, including copper and brass (the copper layer was applied first to enable the brass layer to adhere), galvanising with zinc, phosphating or lacquering. Numerous different composite constructions were also tried, such as having a sheet steel body and a brass base. The brass-coated steel version was the most popular among the ammunition types during the war years.

Propellants

The propellant contained within the 8.8cm Flak shell had as many demands placed upon it as any other component of the shell or gun. Most important of all, it had to be capable of generating the pressures sufficient to deliver the required muzzle velocity dependably, regardless of climate or ambient temperature. It had to be stable in storage, not prone to either degradation or instability – ammunition failing to fire or detonating involuntarily could be twin nightmares for a gun crew. A good propellant also limited smoke and flash, the former obscured the gun crew's visibility of the battlefield, while the latter provided a visible

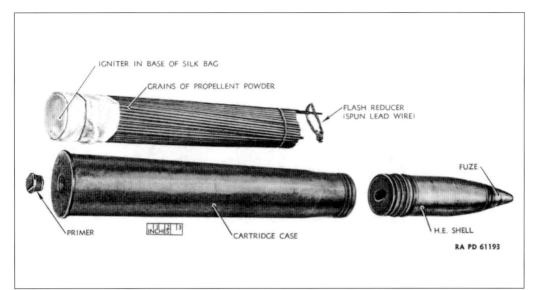

LEFT The internal components of an 8.8cm Sprgr Patr L/4.5, including the 0.87kg (1lb 15oz) change of TNT or amatol, formed into long chains of propellant.
(US War Department)

signature of the gun's location easily identified by the enemy, especially during low-light conditions or in night-time combat. Finally, the nature of a propellant could have a significant impact on the lifespan or performance of a barrel. Propellants that burned with excessive heat had a wearing effect on rifling, shortening the barrel life. A good propellant also limited the amount of fouling deposited along the bore, and thus kept the need for barrel cleaning to a practical minimum.

The two propellants most commonly used in 8.8cm Flak shells were *Diglycolpulver* (abbreviated on shell markings to 'Digl.') and *Gudopulver* (Gu.). Both were double-base propellants (propellants consisted of nitrocellulose but with nitroglycerin or other liquid organic nitrate explosives added). *Diglycolpulver* combined diethyleneglycol dinitrate (DEGN) and nitrocellulose, plus methyl centralite to make the propellant more stable and potassium sulphate to reduce the levels of muzzle flash. *Gudopulver*, by contrast, consisted of *diglycolpulver* and gudol, the latter another flash-reducing agent.

To gain the optimum performance from the propellant, it had to be physically configured in the appropriate way to deliver the right pressures over the right duration. Propellant is formed into 'grains' of a chosen size and shape to control the burn characteristics appropriate to the shell being delivered and the gun delivering it. In the case of the Flak 8.8cm shells, the propellant was formed into long, thin sticks, each almost as long as the shell case. These were held together at the base by a silk bag, which also contained a small gunpowder igniting charge to act as an ignition accelerator between the primer and the propellant. Note also that at the top of the propellant charge, just beneath the base of the warhead, was a length of spun lead wire, which acted as a de-coppering agent within the bore.

One final point about the propellant type for 8.8cm shells was that the chemical composition and the weight of propellant were sometimes altered for service in tropical countries (specifically North Africa), where temperatures exceeded 25°C (77°F). These shells are marked with the letters *Tp.* or *Tropen*, to denote 'tropical'.

BELOW The barrel locking collar of the RA 9 barrel. *Linksgewinde* indicates that the collar has female screw threads; *Los* shows the direction to loosen, *Fest* to tighten.
(Author/Axis Track Services)

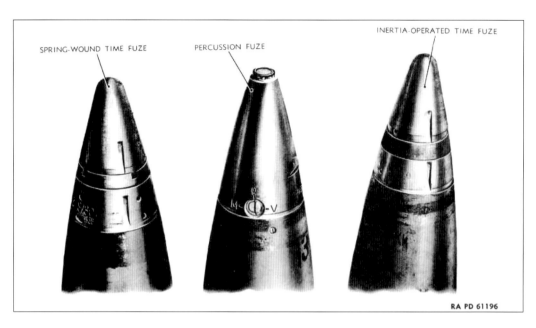

RIGHT Different types of 8.8cm warhead fuze. Both timed and percussion fuzes could be fitted to the HE shell. *(US War Department)*

Fuzes

Before looking at warheads, it is worth spending some time delving into the specifics of German shell fuzes, as used in the 8.8cm Flak ammunition. The focus is warranted because the fuze type was a defining element in how the shell delivered its effects, principally in HE shells, but also in some types of AP ammunition.

We start our analysis by separating German 8.8cm fuzes into two types: time-delay fuzes and percussion fuzes. The names are explanatory. Time-delay fuzes (*Zeitzünder*) allowed the gun team to set the shell to explode at a certain time interval, that interval timed to correspond to a certain range or altitude. Looking internally, the actual mechanisms by which the timing was applied were either a Krupp-Thiel spring-driven mechanical clockwork timer or a Junghans centrifugal drive, the latter indicated by the letters 'Fg' applied to the fuze title, to indicate *Fliegewicht* (flying weight).

The mechanical time-delay fuze most commonly used in 8.8cm Flak ammunition was the Zt Z S/30, an aluminium device measuring 112mm (4.4in) long and with a gauge of 50mm (1.96in). The maximum time setting of this fuze was 30 seconds, giving it enough duration to reach up to the highest altitudes of most contemporary bombers. There were numerous sub-varieties of this fuze – too many to list and analyse here – but the distinctions between them revolved largely around the minimum running time of the fuze (an essential safety feature) of either 1 or 2 seconds, the maximum run time (some went up to 160 seconds), the calibration of the fuze setting (one fuze, for example, detonated the warhead 100m/328ft before the set distance to ensure an effective distribution of fragmentation on the target) and the addition of a percussion fuze. Some of these fuze types were little used in combat. Of the inertia-type

BELOW The gun trunnions on the upper carriage. Note the 8.8cm HE shell in the background, with its gleaming cartridge brass case. *(Author/ Axis Track Services)*

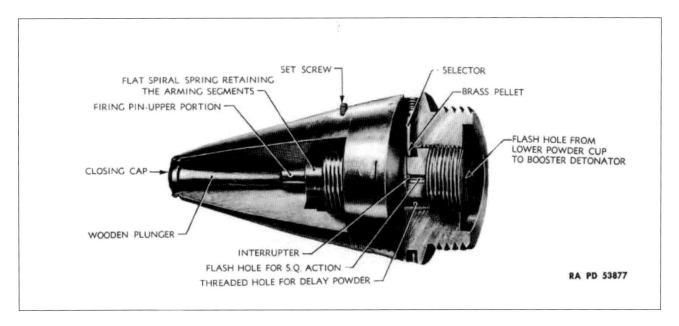

ABOVE The AZ 23 was a nose-mounted percussion fuze, which gave both graze and direct-impact detonation and could be set with a sub-1-second time delay. *(US Army/Air Force)*

fuzes, the core model was the Zt Z S/30 Fg[1], again with a 30-second timer.

The 8.8cm shells did not just rely upon timed fuzes to set them off. A range of percussion fuzes were also fitted, these triggering detonation as they struck the target by either impact (driving a striker or firing pin directly on to the detonator) or by deceleration causing an inertia pellet to slam into a detonator needle. Again, such fuzes could be configured in several different ways to produce varying effects. For a start, the location of the percussion fuze could be either in the nose or in the base. Nose-mounted percussion fuzes (*Aufschlagzünder*) either detonated the explosive content of the warhead the very moment the shell struck the target, or could have an inbuilt 'graze' function, which triggered detonation the second the warhead made contact with any surface, even at an extremely shallow angle. Base-detonating fuzes (*Bodenzünder*), by contrast, were located at the rear of the warhead, which meant that there was a short delay between the initial impact of the warhead and an explosive detonation. Practically, *Bodenzünder* had applications for some types of AP shells and were also perfectly suited to acting as a graze fuze.

One final type of fuze to mention – the 8.8cm Flak chiefly being an AA gun – is the proximity fuze, which triggered warhead detonation when it sensed, via electronic interference components, that an enemy aircraft was within a pre-set distance. Although such fuzes were applied to Allied artillery (AA and field artillery), and they were under development within Germany, they never made it into the warheads of 8.8cm shells, as useful as they would have been.

Before moving on to look at warheads, it is worth reminding ourselves of just how important fuzes were in the grand scheme of things, particularly when it came to the challenge of laying down AA fire. Although AA shells could make a ferociously impressive pattern of shell

LEFT A Luftwaffe gunner on the Eastern Front in 1942 makes a manual adjustment to the fuze settings of an 8.8cm Sprgr Patr L/4.5 high-explosive shell. *(Bundesarchiv, Bild 101I-455-0007-31/ Kamm, Richard/CC-BY-SA 3.0)*

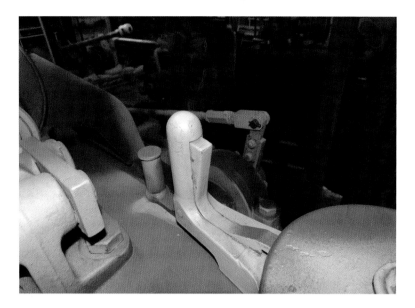

ABOVE A close-up of the handle for the breech actuating mechanism. *(Author/Axis Track Services)*

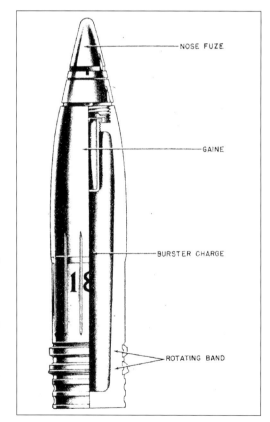

RIGHT This interesting shell is an 8.8cm Sprgr Patr L/4.5 Zt. Z controlled fragmentation round. Note the grooves in the shell body around the bursting charge, which were meant, unsuccessfully, to improve fragmentation. *(US Army/Air Force)*

bursts in the sky, getting them to explode with destructive precision in the direct vicinity of an enemy aircraft was another matter altogether, as Ian Hogg explains:

> The 8.8cm shell, carrying just under 900g (2lb) of explosive, had a lethal burst radius of about 9.14m (30ft): or, in other words, the shell had to burst within 9.14m (30ft) of the aircraft to damage it. Assuming that perfect results were obtained from the predictor, so that the course of the aircraft and the shell coincided, there was one factor that could make nonsense of everything – the fuse. The clockwork fuses used by the Flak gun were probably the best of their kind, but even so they had a tolerance in their timing which can be assumed to be about 0.5% of the flight time. So, assuming that the target was at a height demanding 20sec flight, this half per cent meant that at the speed the shell was travelling the fuse was liable to function anywhere within a 61m (200ft) section of the trajectory, above or below the aircraft, with no control over the precise point of detonation.[16]

Hogg's explanation of the relative imprecision of timed fuzes goes some way to explaining, when taken in the context of all other factors related to accurate AA gunnery, the massive imbalance between shells fired and aircraft hit. There was also the fact that fuzes were prey to all the wear, tear and climatic effects of battlefield use; precision instruments in a very imprecise environment.

Warheads

From fuzes we turn to the warheads themselves. The top-level categories of Flak 8.8cm warheads were HE or AP types. A variety of sub-types fell under each of these categories, each warhead defined by its combination of material composition, shape, explosive content (or not) and fuze configuration. One common feature needs to be noted, however – twin driving bands, situated on the thick end of the warhead just above the case crimping. As with all artillery shells, the driving bands – made of a softer metal than the shell body – serve to engage securely with the rifling, imparting the optimum spin to the shell while also making a gas-tight seal on the warhead, preventing forward gas leakage. In the early days of the Flak series, the driving bands were made of copper, but once demands for this material placed the squeeze on supply, sintered iron was used instead, denoted by the letters 'FES' on the shell.

High-explosive warheads

The 8.8cm HE warheads provided two core tactical functions: (1) Deliver blast and fragmentation effects against personnel, positions and soft-skinned vehicles; (2) Deliver blast and fragmentation effects against aircraft at various altitudes. HE shells were denoted by the abbreviation Sprgr, for *Sprenggranate* (explosive shell), and they could be identified by their paint scheme – yellow, with black stencilled lettering. The basic HE shell was the 8.8cm Sprgr Patr L/4.5, which weighed 9.4kg (20.73lb) and held 0.87kg (1.92lb) of TNT or amatol high explosive. It could be adapted to either AA or field artillery roles by fitting either the AZ 23/28 nose percussion fuze or the Zeit Z S/30 timed fuze. The 8.8cm Sprgr Patr L/4.5 *Gerillt* (meaning 'grooved') was similar in shape, explosive content and fuze options, but it had 15 longitudinal grooves carved along the shell body, ostensibly to improve the fragmentation effects. In this regard it failed, as only internal grooving would have enhanced fragmentation patterns.

An interesting, but largely experimental, addition to the 8.8cm warhead range was the 8.8cm Sch Sprgr Patr L/4.5 incendiary shell. Weighing a little more than the standard HE shells at 10.06kg (22.18lb), this warhead contained an explosive bursting charge that flung out metal shrapnel, each piece holding an incendiary compound of barium nitrate/magnesium. The idea was that the shell would scatter fire-starting shrapnel over a wide area. This shell did not reach significant production, as it was superseded by the 8.8cm Br Sch Sprgr Patr L/4, which followed a similar principle to the former incendiary shell, but with some added complexities. When detonated by a 113g (4oz) bursting charge, the shell dispensed 72 incendiary pellets from its nose section via a 57g (2oz) propelling charge, each pellet containing the incendiary compound but detonated by its own firing pin and detonator. The problem was that getting each pellet to strike the firing pin precisely was more down to luck than engineering, even though a million of these shells were ordered; the ammunition started to enter action in February 1944. The design of the pellets was improved with a more aerodynamic shape that gave a better

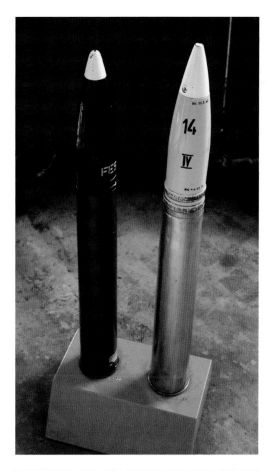

LEFT Armour-piercing (left) and high-explosive 8.8cm rounds set side by side; note the larger warhead dimensions of the HE round. *(Banznerfahrer)*

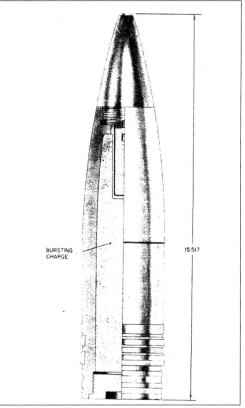

LEFT A cross-sectional diagram of the 8.8cm Sprgr Patr L/4.5 high-explosive shell, showing the TNT or amatol bursting charge and fuzed with either a nose percussion fuze or mechanical time fuze. *(US Army/Air Force)*

RIGHT An 8.8cm high-explosive shell, the type of warhead denoted by the yellow band of colour. This particular vintage shell is fitted with a percussion fuze at the nose. *(Author/Axis Track Services)*

FAR RIGHT The metal container for an individual 8.8cm shell. *(US Army/Air Force)*

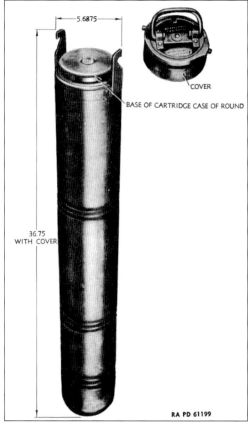

NUMBERS ON GERMAN SHELL INDICATING TYPE OF HE FILLER	
No. on shell	**Type of filler**
1	Fp 02 (TNT) in paper or cardboard container
2	Grf 88 (picric acid) in paper or cardboard container
10	Fp 02 plus Fp 5 plus Fp 10 (TNT fillers) in paper or cardboard container
13	Fp 40/60 (40-60 amatol, poured)
14	Fp 02 (TNT), poured
32	Np 10 (PETN filler) in paraffin-waxed paper wrapping
36/38	Np 40 plus Np 60 (PETN fillers) in paraffin-waxed paper wrapping
91	H 5 (Cyclonite; RDX) in paraffin-waxed paper wrapping

GERMAN EXPLOSIVES, ABBREVIATIONS		
Abbreviation	**German nomenclature**	**English equivalent**
Fp 02	Füllpulver 02	TNT
Fp 5	Füllpulver 5	TNT with 5 per cent montan wax
Fp 10	Füllpulver 10	TNT with 10 per cent montan wax
Fp 40/60	Füllpulver 40/60	40-60 amatol, poured
G of 88	Granatfülling 88	Picric acid
H	Hexagen	Cyclonite, RDX
H5	Hexagen 5	Cyclonite with 5 per cent montan wax
Np	Nitropenta	PETN; penthrite
Np 10	Nitropenta 10	PETN with 10 per cent montan wax
Np 40	Nitropenta 40	PETN with 40 per cent montan wax
Np 65	Nitropenta 65	PETN with 65 per cent montan wax[17]

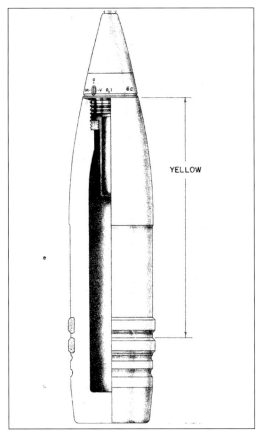

ABOVE The two sintered-iron driving bands on this 8.8cm shell were the only part of the warhead that fully engaged with the rifling during the passage down the bore. *(Author/Axis Track Services)*

LEFT The 8.8cm Sprgr Patr L/4.7 FES, another high-explosive round but with the 'FES' suffix indicating the presence of two sintered-iron driving bands. *(US Army/Air Force)*

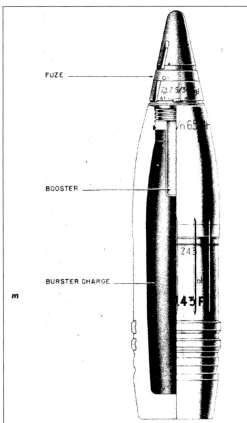

assurance of incendiary detonation, but this version did not see production before the end of the war.

Anti-tank warheads

Flak 8.8cm AT shells, like most anti-armour warheads of the time, had one key goal – penetration of the maximum depth of armour. There were two basic options for accomplishing this. First, you could apply kinetic energy alone via a dense, heavy penetrator warhead, using speed and warhead profile to punch through the armour and unleash destruction inside through 'spall' – massive fragmentation exploding around the interior of the vehicle. An explosive element could be added to such shells, but it was likely to be small in volume and serving a secondary purpose to the penetration. (Having some explosive or incendiary mixture fitted to an AT shell did, however, provide a useful visual marker for the gun crew to see whether they had actually hit a long-range target or not.)

The second route was to use shaped-charge principles. The shaped-charge (or hollow-charge)

FAR LEFT The 8.8cm Sprgr Patr Flak 41 was the high-explosive shell for the Flak 41 AA gun. Note the streamlined base, to reduce drag during flight. *(US Army/Air Force)*

warhead features an open cone of metal, typically copper, seated behind the warhead nose, with the open end of the cone facing forwards (hence the distinctive lozenge profile of many shaped-charge warheads). Explosives are packaged around the outside of the cone (the interior of the cone remains hollow, hence the alternative name). When the shell strikes a target and the fuze detonates the explosive, the blast collapses the metal cone and transforms it into a thin, super-hot (8,000°C/14,432°F) jet of molten metal particles travelling at hypersonic speeds along the central axis of the cone. This jet is capable of cutting through substantial armour.

One of the great virtues of the shaped-charge warhead is that it does not rely upon velocity to achieve its penetration, hence it was the warhead of choice for most low-velocity man-portable AT missile launchers in the Second World War, such as the German *Panzerfaust* and the Allied Bazooka. For rifled AT artillery, however, there are problems. What we today call the High-Explosive Anti-Tank (HEAT) shell – basically a tank version of the hollow charge – works best from smooth-bore guns, because the gyroscopic spin imparted by rifling weakens the formation of an effective molten jet when the shell detonates. This is why the HEAT shells fired from modern smooth-bore tank guns are often fin-stabilised, the fins replacing the stabilising role previously provided by the rifling. A spin-stabilised HEAT round is still a dangerous entity for enemy armour, for sure, but it will not have the same penetration as either a rifled kinetic-energy round or a non-spinning HEAT shell.

With this basic theory in mind, we can look at the specific types of 8.8cm AT shell. The core model was the 8.8cm Pzgr Patr warhead, which weighed a total of 9.5kg (20.95lb). It relied principally upon kinetic energy to do its work, having a chromium molybdenum penetrator plus a mild steel ballistic cap to improve the shell's streamlining and therefore flight characteristics and sustained velocity. It did, however, include a small 155g (5.47oz) explosive charge of TNT/Wax or PETN/Wax, triggered by a base detonating fuze, plus a tracer element so that the gunners could observe the flight of the shell. Against the

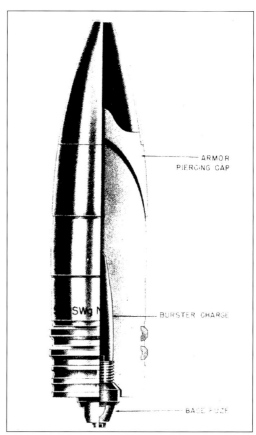

RIGHT The Pzgr 39/43 was an armour-piercing 8.8cm tank/anti-tank gun shell, fitted with a small 0.1kg (0.3lb) bursting charge and a base detonating fuze to give a delay for penetration. *(US Army/Air Force)*

FAR RIGHT Compared to the standard 8.8cm Pzgr Patr, the 8.8cm Pzgr Patr 39 had a slightly re-contoured ballistic cap and iron driving bands. *(US Army/Air Force)*

standard homogeneous steel armour of the time, the 8.8cm Pzgr Patr could achieve a penetration of just over 100mm (3.97in) at a 30-degree angle at 914m (1,000yd), dropping down to around 84mm (3.3in) at double that range (with the same angle of impact).

The 8.8cm Pzgr Patr 39 that superseded the 8.8cm Pzgr Patr had minor differences, specifically the brass driving bands of the previous shell were replaced by sintered-iron versions, the ballistic cap was improved slightly for aerodynamic qualities, and no tracer was included. The 8.8cm Pzgr Patr 40, by contrast, was a new design, centred on a sub-calibre tungsten carbide penetrator. Tungsten carbide offered superior penetrative capabilities and, by having a sub-calibre penetrator, the kinetic energy applied to the round was concentrated against a smaller area. In tests and combat, however, it was shown that the benefits of the new shell were primarily felt at short ranges of under 1,000m (1,094yd) – at 500m (574yd) the Pzgr Patr 40 could penetrate 126mm (4.96in), about 16mm (0.6in) more than the standard steel shot. Yet at 1,000m and beyond the improvements were negligible, largely because the air resistance effects on the shell drained the muzzle velocity. There was also the wider problem of accessing tungsten carbide in the first place, the constituents of which (wolframite and scheelite) were the subject of a vice-like Allied blockade. The manufacture of precision engineering tools took first claim over the raw materials that did make it into the country, hence the production of the Pzgr Patr 40 was

ABOVE The rear carriage of the Flak 37. A shell loading tray could be fitted to the left side of the carriage. *(Author/Axis Track Services)*

BELOW LEFT One of the three-shell 8.8cm ammo crates, seen here in the back of an SdKfz 7 transporter. *(Author/Axis Track Services)*

BELOW The wicker 8.8cm ammo crate with two shells in situ. Rubber stops at the bottom of the case protected the shell fuzes from jarring. *(Author/Axis Track Services)*

severely curtailed, to the extent that the shells were given an emergency reserve status, only to be used in critical situations.

Although, as discussed above, hollow-charge shells were not ideally suited to use in rifled guns, this did not stop the development of such a warhead for the 8.8cm Flak guns. It was known as the 8.8cm H1 Gr Patr 39 Flak L/4.7. Total weight of the shell was just 7.2kg (15lb 13oz), the low weight reflecting the empty cavity within the nose of the shell. The warhead was still to be respected by the enemy, as it could penetrate more than 100mm (39in) of armour, but the penetration figures given on paper did not correspond with some of the feedback coming in from the field. A particularly interesting document in this regard relates to the use of such ammunition in the *Ferdinand/Elefant* tank-destroyers. In February 1943, officials in the Wa Prüf 6 (one of the departments of the Waffenamt) wrote to the Inspector of Armament saying:

The Abteilung *1 understands that all guns which are able to fire PzGr39 and HL-rounds [HL = Hohlladung, shaped charge] will be predominantly provided with shaped charge rounds.*

The department refers with emphasis that this plan does not correspond with the application of the respective high-performance anti-tank guns and tank guns. Until today, the performance of the shaped-charge rounds is too weak when compared to that of AP rounds. For these guns, HEAT rounds can only be a stopgap. Even if we do not underestimate the production problems with AP rounds, we advise by all means to provide these weapons with both types at a ratio of 1:1 AP rounds to HEAT rounds at minimum dependent on production output.[18]

Part of the problem with the HEAT rounds was that they generally had to be fired at lower muzzle velocities to reduce the spin influence on the warhead, thus altering their flight characteristics over range (the gunner would have to adjust for a more arced trajectory). HEAT rounds were used by 8.8cm Flak guns, and they could be armour-killers, but they were often applied to non-armoured targets because they delivered a more explosive effect than the inert penetrator rounds.

One of the other specialist rounds used by the 8.8cm Flak was the 8.8cm *Leuchtgeschoss L./4.4*, which was an illumination shell destined principally for naval service. The 9.5kg (20lb 15oz) shell in this case actually contained a packed parachute flare, and it was fitted with a timed Zt Z S/60nA fuze plus a small black-powder charge. When the shell was fired by the round's 2.09kg (4lb 10oz) propelling charge, at the correct altitude (based on the fuze settings) the black-powder charge would go off, blowing the base off the projectile and allowing the parachute flare to fall out, the flare itself burning for 23 seconds at an output of 375,000 candlepower.

Most of the shells considered so far have related to the Flak 18/36/37 weapons. The Flak 41 gun's ammunition was naturally related to such types, but with the requisite adaptations for achieving the gun's higher muzzle velocities. Its chief HE shell was the 8.8cm Sprgr Patr Flak 41. Here the shell acquired a slightly longer and more streamlined profile – the base was given a more 'boat-tail' design to reduce drag in flight –

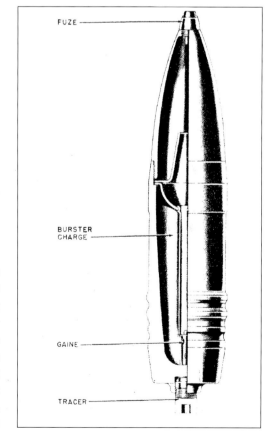

RIGHT The shell here is an 8.8cm Gr Patr 39 HL, the 'HL' referring to *Hohlladung*, or 'hollow charge'. We can see empty space in front of the metal liner at the front of the shell. *(US Army/Air Force)*

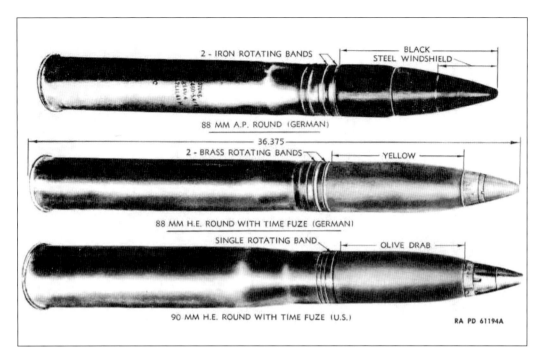

LEFT The US Technical Manual on the 8.8cm Flak gun here compares the standard 8.8cm armour-piercing (top) and high-explosive (centre) shells with the American 90mm HE round. Note the ballistic cap on the AP warhead. *(US War Department)*

although the explosive content was essentially the same as that of the L/4.5. The big difference was in the shell case length and contents. Whereas the L/4.5 had a case length of 570mm (22.44in) and a propelling charge of 2.41kg (5lb 5oz) of Digl. R P, the Flak 41 case measured 859mm (33.82in) and held 5.12kg (11lb 5oz) of Gudol R P. This additional propellant force gave the Flak 41 the extra power that was desired. The shell could be fitted with either impact or timed fuzes. A grooved version of the round was also produced – the 8.8cm Sprgr Patr Flak 41 *Gerillt*.

For its AT rounds, the Flak 41 fired the same warhead type as the other Flak guns, but again with adaptations to the shell case and the propelling charge. The 8.8cm Pzgr Patr 39/1 Flak 49 combined the standard Pzgr Patr 39 shell with the extended Flak 41 shell case and a 5.42kg (11lb 15oz) propelling charge, to generate a muzzle velocity of 980m/sec (3,215ft/sec). The 8.8cm Pzgr Patr Flak 41 was the same configuration but with a heavier propelling charge that delivered 1,150m/sec (3,773ft/sec).

Pak 43 and tank gun ammunition

The relationship between the ammunition types used by the Flak 18/36/37/41 guns and those chambered by the 8.8cm AT/tank weapons was naturally close, albeit with some differences worthy of note. Looking at the Pak 43 first, the ammunition for this weapon closely followed some of the AT types already discussed, with some modifications both subtle

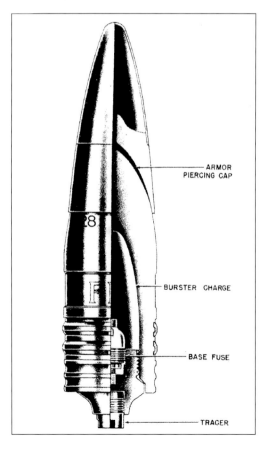

LEFT An 8.8cm Pzgr Patr anti-tank round, fuzed with the Bd f base-detonating fuze. The base-mounted tracer enabled the gunners to observe better the impact of the round. *(US Army/Air Force)*

RIGHT The 8.8cm Pzgr 39/1 was fired from the Pak 43 plus the 8.8cm tank and tank-destroyer guns. Total weight of the shell was 10.2kg (22.5lb). *(US Army/Air Force)*

FAR RIGHT The 8.8cm Sprgr Patr 43 was a high-explosive round used by the KwK 43 tank gun and the Pak 43 anti-tank gun. *(US Army/Air Force)*

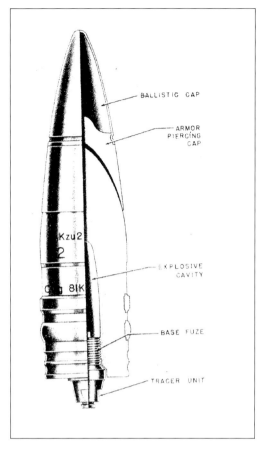

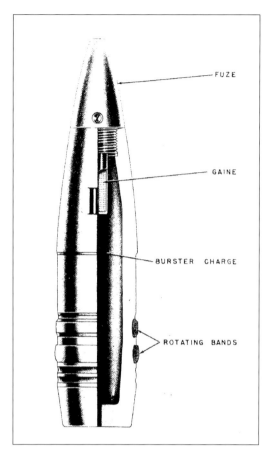

and significant. As a general rule, the Pak 43 principally fired shells with percussion fuzes – shells with timed fuzes were better handled by the AA and field artillery, which engaged a spectrum of target types with indirect and predicted fire, whereas the Pak 43 was almost exclusively used for direct fire against 'hard' armoured targets. Second, the Pak 43 shells were modified to use electrical ignition, the technology of which is explained above.

The basic type of armour-piercing shell for the Pak 43 was the 8.8cm Pzgr Patr 39/1. Apart from the C/22 electrical ignition system, the only major difference between this shell and the standard Pzgr Patr 39 was that the Pak 43 version was fuzed slightly differently; it used the Bd Z 5127 base-detonating fuze, which gave a very fast detonation time. The 8.8cm Pzgr Patr 39/43 was almost identical to the 39/1, but with the exception that the sintered-iron driving bands were slightly wider (at 16.5mm/0.6in) and heavier than those of its predecessor. This modification was made in light of the fact that after the Pak 43 had fired more than 500 rounds, wear inside the bore meant that the standard shells lost their precision fit and therefore their accuracy. By expanding the driving bands just a fraction, however, the Pak 43 could exercise its full barrel life of 2,000 rounds.

Another armour-piercing shell used by the Pak 43 was the 8.8cm Pzgr Patr 40, which fired a tungsten carbide round. Unlike the sub-calibre projectile used for the Flak guns, this round had a full-calibre warhead, weighing 7.3kg (16lb 1oz). Like the Flak guns, however, the limitations of Germany's access to the materials for producing tungsten carbide meant that this shell was not in common use. Pak 43s also had their own hollow-charge shells: the 8.8cm Gr Patr 39 H1 and the modified version with wider driving bands for post-500-shot guns – the Gr Patr 39/43 H1. To maintain lower velocities, these projectiles, which weighed 7.65kg (16lb 14oz), had a propelling charge of 1.7kg (3lb 12oz) of Gudol R P, as opposed to the 6.83kg (15lb) in the 39/1 shell. Thus the muzzle velocity dropped to around 600m/sec (1,968ft/sec), with the book-figure penetration at 1,000m (30 degrees impact) given as 90mm (3.54in).

Two types of HE shell completed the ammunition profile for the Pak 43, again with the distinction between the two relating to the width and weight of the driving bands. The 8.8cm Sprgr Patr L/4.7 fired the same 9.4kg (20.73lb) projectile as the Flak guns, but purely with the AZ 23/28 nose-mounted percussion fuze. Variations in the propelling charge and the driving bands made the complete round somewhat heavier than the anti-aircraft versions, at 19.3kg (42.56lb). The 8.8cm Sprgr Patr 43 was the HE round with the wider driving bands, although this round weighed the same as the L/4.7.

When looking at the KwK 43, mounted in various tanks and SP guns, the ammunition picture is simplified by the fact that it basically used the same ammunition as the Pak 43, including the versions with the wider driving bands. For once, German ordnance technicians restrained themselves from any further proliferation of types, which was rather atypical of them.

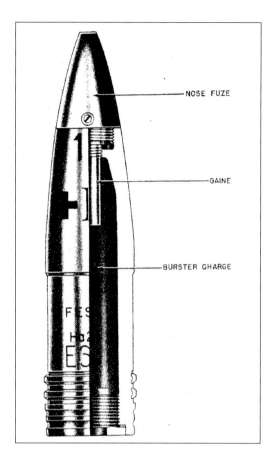

LEFT The 8.8cm Sprgr Patr L/4.5 high-explosive shell; the 'gaine' is an explosive charge linking the fuze with the main warhead charge. *(US Army/Air Force)*

BELOW The crew of a Tiger I tank carefully pass shells inside for storage. An impressive 92 shells could be stored in the interior of the tank. *(Bundesarchiv, Bild 101I-022-2948-28/Wolff, Paul Dr./CC-BY-SA 3.0)*

Chapter Four

Crew roles and gun operation

The 8.8cm Flak guns were sophisticated and complex pieces of equipment. Being able to deploy and operate such weapons under combat conditions required a crew that was exceptionally well drilled in every aspect of the gun's function, able to perform the core actions almost instinctively.

OPPOSITE An 8.8cm gun was the centrepiece of a small world in which the gun crew would rapidly develop the closeness of family. This crew and their 88 are in France in the spring of 1940 where they are shelling the Maginot Line. *(Bundesarchiv, Bild 101I-769-0231-11)*

ABOVE A very grainy image shows the gun crew of the Flak 36 named *Emil*, illustrating how the crew might personalise their weapons. *(Heinz Radtke/Family Archive Norbert Radtke)*

There is always a problem in talking about the roles, responsibilities and duties of military personnel in times of conflict. This is the issue of the gap between theory and reality, which could be brutally wide in wartime. While for both writer and reader it gives comforting precision – and indeed an essential framework – to quote from the 'textbook' sources about how things should ideally be done, in action the rules might dissolve significantly. For example, if your 11-man gun crew has been whittled down to just four or five on account of dysentery sweeping through the battery, textbook procedures will doubtless have to be improvised to meet reality.

The following account is a good example of how external forces affected the procedures of operating the 8.8cm Flak gun in the field, even to the level of basic manual dexterity. It comes from one Kanonier Wolfgang Kirst, serving on the Eastern Front during one of the brutal winters:[19]

No building anywhere in which we could have found shelter from the cold. Far and wide only empty fields. We spent the night on the open road at minus 32 degrees. Toward morning, an order from the unit reached us: 'Battery goes into position immediately for anti-aircraft fire.' In the shortest time we had to be ready to fire, and in this cold! The tracks of the heavy telling tractors skidded over the ice-slick ground. They did not catch hold. The gun stayed put. A second towing tractor was hooked on. The heavy gun moved a little. Four gun crews pushed or pulled on long ropes. Thus we moved the gun into position. Finally we moved to the second, then the third and fourth. The command device followed.

Take position, level the gun. Lay out the communication cable, fit the cable couplings

RIGHT Here, stored atop the mudguard and gun winch mechanism, are basic tools of the trade – a length of thick rope and a pickaxe, the latter particularly useful for cutting ice-hardened earth for the carriage-levelling jack feet. *(Author/Axis Track Services)*

together. It was 35 degrees below zero. Here a piece sticks, there one didn't fit into the other as intended. Take off your gloves! With bare hands, iron parts had to be handled. Our fingers froze to the metal at once, sticking as fast as if they were baked on. Bits of skin tore off. The men had to operate the command device without steel helmets. The man with the rangefinder did not have it easy. Every breath fogged the viewer lenses, taking his view away. We had learned what corrections had to be made when shooting at temperatures like this, we knew all the things that had to be taken into consideration. So this time too, the batteries were ready to fire when the Soviets attacked from the air.

The shells whistled up at the bombers. The trajectories of the shells could be seen as fine white streaks of condensation in the air. Several planes were shot down by our direct hits; the others turned around and went off into the distance.

We had no time to blow our noses, for we had to convert to ground combat quickly. Once again the battery was divided into separate Flak battle troops for anti-tank action. After three hours our guns lay in wait several kilometers farther on.

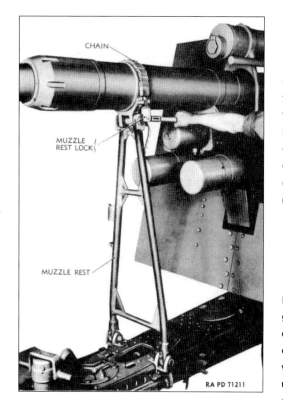

LEFT Even in the height of battle, the crew had to avoid driving away with the gun's muzzle not fixed by the muzzle rest lock. Doing so could result in severe damage to elevation and traverse gears. *(US War Department)*

As expected, the tanks attack in the first light of morning. Before the blinding-white brightness of the snow-covered landscape they rose up as dark points at a great distance. 'Let them advance calmly!' the

BELOW A Luftwaffe gun crew in Russia, during the months of more amenable weather. Note the use of grasses to camouflage both gun and position. *(Bundesarchiv, Bild 101I-455-0006-03/ Kamm, Richard/CC-BY-SA 3.0)*

ABOVE Being an 8.8cm gunner meant a lot of hard labour. Here troops in Berlin-Karow dig another anti-aircraft gun position, adding to those behind them. *(Heinz Radtke/Family Archive Norbert Radtke)*

lieutenant called to his gun leader, then he stamped through the deep snow to the second gun. The tanks came on. The front one turned broadside to us. 'Fire!' The Colossus rolled on, heading for a house, broke into the cottage, stopped there, exploded. The house collapsed.

A second tank came toward the gun next to ours, up to 100 meters. The lieutenant ran over to the gun. We heard the impact of a shell and saw him get up and go on behind a cloud of smoke and dirt. Now he was at the 88 and shoved a shell in, aimed at the second tank and fired. The shot went off. At the same moment, it must have been in the same fraction of a second, muzzle fire sprang up. The enemy seen by the man at the gun flew all over the place. Their detonations ripped through the air in a single mighty explosion.

The inertia straining against every action here is extreme: it is physically punishing just to tow the gun into position, handle its metal parts and even simply to look through the optical sights. And yet, through all the evident frustration, the gun crew still manage to bring their weapon into action and start destroying enemy vehicles. Such exploits are only possible when standards of training have been high in the first place, enabling the crew to maintain some level of functionality in the midst of chaos.

Yet we should be careful about ascribing proficiency to every German Flak team. The Flak crews on the combat front lines, particularly on the Eastern Front, were often some of the best artillery personnel, naturally sent to where it appeared they were needed the most. The picture of AA crew back in Germany itself could be very different as the war progressed, despite the fact that this was where the bulk of the Flak guns were in operation against a devastating and unrelenting strategic bombing campaign. In 1943 the total number of troops employed in AA work numbered more than a million, if we include auxiliary staff. But from this pot of personnel the Wehrmacht would draw out large numbers, typically the better trained or physically fitter, and channel them out into either other roles in the armed forces or into front-line artillery roles. Thus it was that the Flak defence of the Reich became acutely dependent upon auxiliaries, as Boog, Krebs and Vogel explain with some striking figures:[20]

By November 1943 more than 400,000 men and women were working in flak defence as auxiliaries; including 80,000 schoolboys as 'flak helpers', 64,000 RAD men, 4,500 Hiwis, 60,000 PoWs, and 176,000 men in the home-guard flak batteries. Attempts to persuade western Europeans to volunteer as flak auxiliaries in Germany were quickly abandoned as hopeless.

The 'Luftwaffe-' or 'flak-helpers' (Flakhelfer) were schoolboy soldiers who, besides being trained and used to serve the guns on flak sites, continued to have lessons from their teachers; their ages ranged from 15 to 17. The first of them were enrolled on 15 February 1943, as the result of a Führer decision of 7 January. As a rule they lived up to expectations, and generally made up in keenness for what they lacked in training. The Reich Labour Service provided 420 new flak batteries in Germany. One protecting the Politz hydrogenation plant at Stettin, to take one example, mustered one officer and 10–15 only partially fit soldiers. Some NCO posts could not be filled. The gun crews consisted of 15- to 16-year-old 'flak helpers'; 18- to 19-year-old women auxiliaries operated the predictors. The shells were manhandled by some 20

Russian PoW Hiwis, headed by a Russian sergeant, who were guarded by the 'flak helpers' and brought to the guns when the air-raid warning was sounded. Searchlights were crewed by 17- to 18-year-old 'RAD girls' under the charge of a technical NCO. With such a mixture of personnel, whose members were frequently changing, it was impossible to ensure a peacetime level of training for the home-guard flak units.

This astonishing profile of exactly who might be classified as a 'combatant' in wartime illustrates our opening point about the gap between theory and reality. This being said, deficiencies in pre-service training could be compensated for to a large degree by regular in-theatre drills, designed to instil actions such as target acquisition, fuze setting and loading at the level of instinct. Official manual advice notes how drill rounds should be used regularly to practise the loading procedures, and that if actual aircraft were not available for sighting drills, then small balloons could be released in the distance to test spotting. A wise gun leader would ensure that his crew conducted frequent drills and exercises, even in wartime settings, and especially if new recruits or replacements joined the team.

In this chapter, we look more closely at the way 8.8cm Flak crews were meant to handle their weapons in peace and in war. Here we mostly focus on the theoretical principles of gun handling, as without understanding them we cannot interpret the ways that they might be adapted in wartime, the main focus of our next two chapters.

ABOVE This artwork illustrates the sheer amount of labour that went into creating an anti-aircraft position. The wood-supported earth banks around the gun would protect from bomb fragments. *(AirSeaLand Photos/ Cody Images)*

LEFT Not all 88 crews were German. Here we see a Flak 37 used by Slovak insurgents during the national uprising of 1944. *(Pavel Pelech)*

ABOVE Another scene of a Flak crew at rest, this time in France in the summer of 1944. The Luftwaffe soldiers wear camouflaged combat overalls. *(Bundesarchiv, Bild 101I-496-3469-24/ Zwirner/CC-BY-SA 3.0)*

Manning the 8.8cm Flak

At full complement and in accordance with pre-war training manuals, the gun crew of an 8.8cm Flak would consist of 12 men. These were the gun leader, nine *Kanoniere* (cannoniers) to perform all the physical actions of setting up, loading, aiming and firing the gun, plus one or two cannoniers assigned to the ammunition staff, whose duty it was to keep the gun team in shells sufficient to meet demand. This whole team would be carried in the towing truck.

To ensure that each *Kanonier* knew his specific role, he would be given a number – K1, K2, etc. – although his number would often change as he was cross-trained in different duties around the gun. During training, the gun crew would learn very regimented procedures, governed by their individual numbers. For example, inside the truck there were two rows of seating. In the driver's compartment at the very front, there were the vehicle leader in the driver's seat, the vehicle attendant (essentially the driver's assistant) in the middle seat, and the gun leader on the right. In the rear compartment, eight *Kanoniere* would be seated in two rows of four, K1–K4 in the seating nearest to the cab, while K5–K8 sat near to the tailgate. This seating arrangement allowed for fluid mounting and dismounting. Following dismounting (at least in orderly peacetime, parade ground or training conditions), the gun crew would essentially stand to the rear of the gun and line up in a way that reflected the seated positions within the truck. When getting back on to the truck, the procedure would be equally regimented, with K1, 2, 5, 6 and 9 entering the truck from the right-hand side of the vehicle, and K3, 4, 7 and 8 from the left.

Apart from the techniques of firing the gun at a live target, probably the most important duty performed by the gun crew was to set the gun up ready for action from its towed state. Each crew member would be drilled constantly in this procedure, and the crew that could bring the gun into action within 2 minutes had achieved a polished proficiency. The speed with which the set-up was performed depended

ABOVE Framed by a beautiful vista of southern France, a Flak team manhandle one of the carriage bogies to a coastal position. *(Bundesarchiv, Bild 101I-258-1324-12/ Micheljack/CC-BY-SA 3.0)*

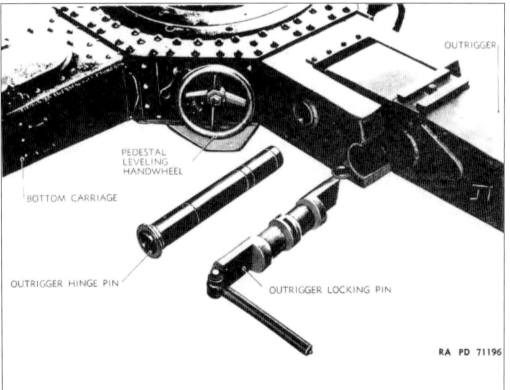

LEFT A diagram of the hinge and locking mechanism for the 8.8cm Flak outriggers. *(US War Department)*

CREW ROLES AND GUN OPERATION

heavily on a clockwork-like integration of the whole team, each man working around the gun in a coordinated fashion. The following passage from the US Army's manual on the 8.8cm Flak gives a sense of the complexity involved just in transferring the gun from its wheels to its platform, while also providing a useful guide to this critical procedure:[21]

21. TO PLACE THE WEAPON IN FIRING POSITION.

a. *The piece may be fired from the wheels but must be emplaced for high angle fire. To fire from the wheels:*
   (1) *Unlimber the prime mover from the drawbar.*
   (2) *Set the hand brakes on the rear bogie.*
   (3) *Engage the elevating gear clutch by pulling the clutch lever to its downward position.*
   (4) *Release the muzzle rest by:*
      (a) *Unscrewing the muzzle rest lock so that the chain may be swung over the barrel.*
      (b) *Elevating the gun slightly so that the muzzle rest may be pushed forward and down onto the bottom carriage.*
   (5) *Level the top carriage by cross-leveling handwheels, using the level indicator for reference.*
   (6) *Unfold the rammer guard to the operating position.*
   (7) *Cock the rammer assembly by rotating the rammer crank handle in a counterclockwise direction.*

h. *To Emplace the Mount.*
   (1) *Unlimber the prime mover from the drawbar.*
      NOTE: The operation of disconnecting both bogies is identical.
   (2) *Operate the winch until the chain takes all the weight from the locking jaws.*
   (3) *While one man steadies the winch, disengage one locking jaw at a time by raising handle. Repeat for the other locking jaw on the bogie.*
   (4) *Simultaneously with the above, lower the side outriggers by performing the following steps. NOTE: The instruction plate on the left outrigger reads 'VOR AUSLOSEN*

ABOVE Emplacing the gun. The winch takes the gun's weight of the bogie locking jaws, and the gunners then release the locking jaws using the handles. *(US War Department)*

RIGHT Releasing the locking bar plunger and the safety chain allowed the outrigger to be swung down into its support position. *(US War Department)*

DER STUTZEN SEITENHOLM DURCH 2 MANN FESTHALTEN,' which translated means, 'Before releasing the side outrigger supports, steady (the outriggers) by 2 men.'
  (a) Releasing the locking bar plunger.
  (b) Removing the safety chains.
  (c) Swinging the outriggers down.
  (d) Locking the outriggers in place by rotating the locking pins.
(5) When the mount is completely lowered, unhook the bogie chains.
(6) Disengage the hooks securing the bogies to the mount. NOTE: Unhook the front bogie first.
(7) Remove the bogies, connect them together with the transporter bar, and wheel them away as a complete trailer unit.
(8) Engage the elevating gear clutch by pulling the clutch lever to its downward position.
(9) Release the muzzle rest by:
  (a) Unscrewing the muzzle rest lock so that the chain may be swung over the barrel.
  (b) Elevating the gun slightly so that the muzzle rest may be pushed forward and down onto the bottom carriage.
(10) Support the bottom carriage and side outriggers with the leveling jacks.
(11) Secure the mount in position by driving the stakes through the bottom carriage and outriggers.
(12) Level the top carriage by the cross-leveling handwheels, using the level indicator for reference.
(13) Unfold the rammer guard to the operating position.
(14) Cock the rammer assembly by rotating the rammer crank handle in a counterclockwise direction.

Note that the procedure here does not include setting up the connections and relationships between the fire-control team, gun and the director equipment. For fixed AA positions within the Reich, such activities could be performed at reasonable leisure, but in mobile front-line war zones the speed of these motions could make the difference between life and death.

ABOVE The towing connection of the Flak 8.8cm's rear bogie; a solid metal towing bar connected vehicle and bogie. *(Author/Muckleburgh Collection)*

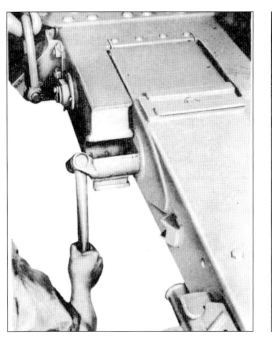

FAR LEFT Locking the outriggers into place by rotating the outrigger locking pin. *(US War Department)*

LEFT This US soldier observes the level indicator as he turns the cross-levelling handwheel to get the gun carriage perfectly horizontal. *(US War Department)*

RIGHT, BELOW AND BOTTOM The ZF 3×8 Flak was an optional sight fitting for quick direct fire against enemy ground targets. These sights could not be used effectively against aerial targets. *(Author/Axis Track Services)*

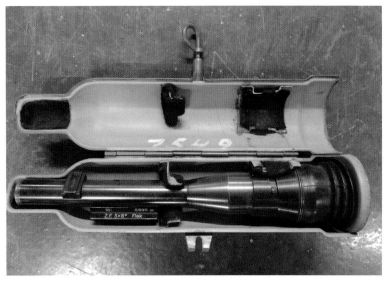

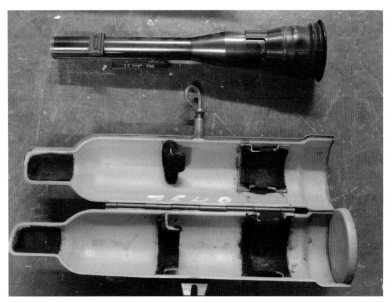

Handling the gun in action

In action, the gun crew had to work together with intent to perform multiple actions either in a near-simultaneous execution or at least a smooth and seamless flow. Prior to firing, the gun had to be aligned with the zeroed aiming circle device, to ensure that the weapon's barrel was properly oriented in accordance with the fire-control network. Using the mils displays on their weapon sights, the gunners also ensured that each gun in a battery was similarly aligned. Once a particular gun was aligned to the 'zero direction', then the gun leader would announce 'Gun X – ready to fire!'

The wartime *Handbuch für de Flakartilleristen (Der Kanonier) Waffen und Ausbildung der Flakbatterie (8.8cm Flak und 2cm Flak) – Handbook for the Flak Artillery Crew (The Cannonier) Weapons and Instruction of the Flak Battery (8.8cm Flak and 2cm Flak)*, written by Ernst Neumann and published in 1941, provides a short and concise catalogue of operational instruction to trainee Flak crews. The date of its publication meant that it would have incorporated field experience acquired in action from the Spanish Civil War to the first year of the North Africa campaign. Taking one of its combat sections – that concerning engaging aircraft in indirect (predicted) firing, the manual illustrates perfectly how the team had to orient themselves around the gun. In this scenario, the gun leader stood just off to the

LEFT Here the gun leader (right foreground) is receiving verbal fire-direction instructions through his headphones, transferring them to the gunlayers. Note the number of men waiting to supply the gun with shells. *(Heinz Radtke/Family Archive Norbert Radtke)*

RIGHT A German wartime artwork showing a Flak gun in transit, heading towards an area being bombed. The tow vehicle looks like a wheeled truck, rather than a half-track. *(AirSeaLand Photos/Cody Images)*

BELOW Many 8.8cm gunners would travel across Europe as the fortunes of war shifted. Here a US soldier inspects a truck-mounted Flak 37 in Budapest, Hungary. *(AirSeaLand Photos/Cody Images)*

BELOW RIGHT An 8.8cm gun is positioned by a railway siding near Dramburg, Pomerania, in November 1943, waiting to ambush overflying aircraft. *(Heinz Radtke/Family Archive Norbert Radtke)*

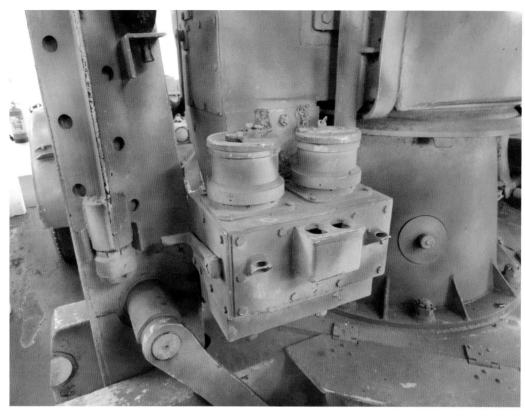

RIGHT The fuze-setting mechanism. Two shell fuzes could be set at any one time. Note the setting crank handle on the side of the box. *(Author/Muckleburgh Collection)*

BELOW Here we look at the rear side of Receiver C, set just above the fuze-setting mechanism, and the electrical connections feeding into it. *(Author/Muckleburgh Collection)*

side of the 8.8cm Flak, monitoring the overall function of the team and the weapon and giving a centralised figure for command.

Four individuals worked directly on the gun itself. K1 would operate the elevation mechanism, aligning the three indicators with the light pattern (on the early model of the receiver) transmitted to the gun from the fire director. Similarly, K2 would operate the traverse mechanism, again moving the receiver pointers to match the lights. The two other men on the gun would be responsible for loading and fuze setting. K7 stood beside the fuze setter, dropping the shells nose-down forcefully to engage with the fuze-setting mechanism. K6 would then actually operate the fuze-setting wheel, aligning the three indicators with the lighted lamps on the fuze-setting receiver by continuously moving the crank handle. Once the shell fuzes were set, K7 could then remove the shells from the setter and load them into the breech for firing. Meanwhile, K4, K5, K8 and K9 arranged themselves in a roughly diagonal pattern out from the gun in support of K7; these four individuals passed shells forward in readiness for loading, with K4 and K5 placing the shells on a mat next to the fuze-setting mechanism or passing the shells to K7 as required.

If there was no physical link with the director, then the procedure was modified to utilise telephone-transmitted information from the fire-control centre. In this situation, the gun leader would announce 'Shoot with command equipment – by telephone transmission.' K7 would give out headset equipment to K1 and a

LEFT One of the three radio headsets stored with the gun, and used when taking fire direction over the telephone, rather than from the fire director. *(Author/Axis Track Services)*

headset and receiver equipment to K6, meaning that all the key players required for setting elevation, traverse and fuze timings were now in the audio communications loop, once they had connected their headset leads. The manual explains that:

> K1 continuously sets and maintains the correct barrel elevation by turning the elevation mechanism so that the barrel indicator matches the commanded number on the elevation arc scale.
> K2 continuously keeps the correct barrel traverse direction set by moving the traverse mechanism so that the mark on the scale ring on the pivot mount matches the commanded number.
> K3 continuously keeps the correct fuse setting by moving the crank to match the commanded scale number on the scale in the fuse receiver window.[22]

Neumann's 1941 manual also provides instructions for direct-fire mode, both with and without the fuze setter. While this would have applications for some types of AA fire, it would obviously have more relevance for AT fire. In the case of direct fire, the gun crew immediately rearranged its priorities, with K2 manning the traverse mechanism and looking for the target with the optical searching sight, K1 manning the elevation mechanism, K6 on the fuze setter (if operable) and everyone else (initially) assisting K2 in looking for the enemy target. The gun leader and K2 might subsequently

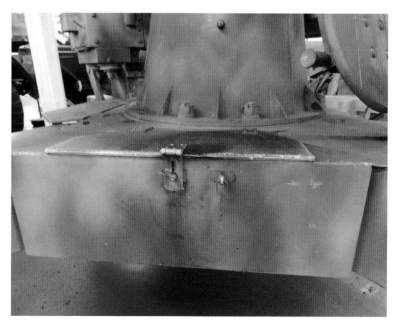

BELOW The pedestal base included a storage compartment, the top of which was also reinforced as a standing platform. *(Author/Axis Track Services)*

RIGHT Here we see to good effect the triple-ring lighting system of the *Übertragungsgerät 30* in the Flak 18 elevation receiver. The three indexes hint at the system's complexity. (US War Department)

receive a locating instruction for the target from an observer or command post, given in the sequence: elevation degree – scale ring number – target elevation (or distance). For example: '45 degrees – 1430 – target elevation 3,000 (thirty hundred) [metres]'. Using such information, K2 would then seek to acquire the target in his telescopic sight, holding the target image so that it was sitting on the horizontal line of the sight reticle, and maintaining visual contact. If all was ready with the gun and its crew, the team could then proceed to fire the weapon, the signal usually being a 3-second ring on the fire bell, the raising of a fire flag or a simple verbal command from the gun leader.

Much of the 1941 manual advice is given over to considerations of engaging enemy aircraft, in accordance with the original intended purpose for the 8.8cm Flak. There is a short section, however, explicitly on firing against ground targets. The manual's author makes one particularly interesting opening point about AT work, in that the gun leader has a responsibility for locating the gun where the prevailing winds will blow the gun smoke from the muzzle away from the fighting position, rather than back over the crew, 'so that he can well observe the target in an unhindered fashion, and so that his commands for the crew can be well understood'. The comment places something of a question mark over the supposed

RIGHT A close-up of the 8.8cm Flak's fuze-setting mechanism. With the shell warhead in the fuze cup, the time-setting handwheel was turned to black out the illuminated lights, which lit up according to altitude date sent from the fire-direction centre. (US War Department)

RIGHT Another view of the elevation quadrant. Note how the gauge is clearly marked from −3 to +85 degrees. *(Author/Axis Track Services)*

'smokeless' qualities of the propellant, although to be fair the smoke generated by firing any gun can partly be attributed to other factors, such as burning gun oil.

When an enemy vehicle was spotted, the gun leader would give his command in the

BELOW A good view of the right side of the Flak 37. Note the wooden gunner's seat folded down in the stored position against the pedestal. *(Author/Muckleburgh Collection)*

RIGHT The traversing limit indicator prevented the traversing wheel operator from making more than two 360-degree turns. *(US War Department)*

FAR RIGHT This image shows a gunner operating the elevation gear clutch, which had to be engaged before elevation could be applied to the gun. *(US War Department)*

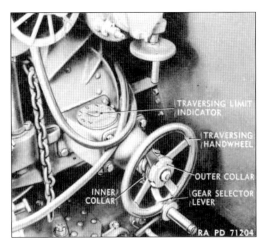

following fashion: Shell type – Target description – Elevation angle – Sight angle, e.g. 'Armour-piercing! Tank crossing summit of hill. 4 degrees 6! Right 10! Fire!' In recognition that anti-armour fire could be a frantic business, especially with enemy armour bearing down on the gun crew at speed, the manual explains that 'If K2 does not find the target according to the target description, the gun leader jumps into his place and, by his own sighting, points out the target to K2 exactly.'

Theatre challenges

The Flak 8.8cm gun crew would live and breathe their weapon on operations. Not only would they spend their waking hours maintaining the gun, or fighting with it, but they would also sleep in its proximity in fighting positions or trenches, the scent of oil and spent propellant in their nostrils at all times. They would also have to do the substantial work involved with emplacing the gun for action, whether in a dug-in and camouflaged fighting position or in a major Berlin Flak emplacement, fully networked to the local and regional fire-control system.

In the following two chapters we look in detail at the 8.8cm Flak's operational record

BELOW The mountainous terrain of Italy presented 88 crews with some enormous fields of fire, but also the problems that came with manoeuvring the weapon around precipitous landscapes. *(Bundesarchiv, Bild 101I-570-1604-18A/Appe [Arppe]/CC-BY-SA 3.0)*

both as an AA weapon and an AT gun, but here it is worth exploring some of the physical and engineering challenges faced by the Flak 8.8cm crews in their more taxing environments, specifically the deserts of North Africa and the winter landscapes of the Eastern Front. Each of these contexts radically changed the nature not only of the tactics used (again the subject of our following chapters), but also on how the gun was physically hauled, emplaced, maintained and fired.

Desert operations

Operations in the North African desert regions stretched the Flak 8.8cm gun crews on physiological, tactical, logistical and mechanical levels. The desert terrain being what it is – flat and arid under direct tropical sunlight – the Flak teams suffered the hardships of all German troops in the theatre, sitting for endless baking hours in emplaced positions scratched out in the desert earth and sand. Camouflage netting was an essential addition to any emplaced position, in an attempt to guard the location from enemy aerial reconnaissance, and the overhead cover, plus the radiant heat of a sun-warmed gun, could make life in such positions almost unbearable. The gun leaders had to be diligent in implementing hydration policies, based on a 'regular and often' approach to drinking (often made problematic by the strain on water logistics in the desert) and for watching out for the onset of heatstroke or severe sunburn. Disease was rife, particularly dysenteric illnesses transferred by flies, which moved rapidly between the crude desert latrine systems and food supplies. Because of this, it was a common occurrence that a gun crew would be reduced in numbers, even without the addition of combat casualties. The Flak teams would have had to become proficient at operating their weapons with skeleton crews.

The severe heat of the desert affected the operation of the 8.8cm gun in some very direct ways. For a start, the gun metal could become startlingly hot under the desert sun, making it painful or even skin-blistering to touch the barrels, breeches and even crucial mechanisms such as the elevation and traverse handles. The heat did not cause damage to the gun itself – the temperatures generated by firing were far more extreme than ambient tropical temperatures – but it added to the misery of the artillery crew's life. One area in which heat did disturb the tactical operation of the 8.8cm gun was in calculating distances using simple optical sights. During the scorching daylight hours from mid-morning until late afternoon, heat haze often obscured the target view for the gunners, while the featureless nature of the terrain meant that it was hard to make range judgements in the absence of measurable landmarks. Furthermore, it was often problematic arriving at a conclusion about exactly what you were looking at. Shortly after the war, Major-General Alfred Toppe, with the assistance of former German theatre commanders from North Africa, wrote an extensive monograph for the US Army explaining the challenges of desert warfare: *Desert Warfare: German Experiences in World*

ABOVE Fitting a muzzle cover, such as this canvas version, was essential to protect the bore from snow, rain, ice, dust and dirt between firing. *(Author/Axis Track Services)*

ABOVE **North Africa, June 1942. A crane assists a gun crew in installing a new barrel on their 8.8cm gun, which appears to be positioned for airfield defence.** *(Bundesarchiv, Bild 101I-443-1587-09A/Zwilling, Ernst A./ CC-BY-SA 3.0)*

BELOW **This artwork gives an impression of the swirling sandy dust in the desert theatre, dust that ingrained itself into every moving part of the gun.** *(AirSeaLand Photos/Cody Images)*

War II. (The work was written in its fullest extent in 1952, but such was its value that the work was published in an edited-down format in 1991, from which we quote here.) In this book, he offered the following insight into the effects of heat and light upon observation:

> The intensive sunlight had a dazzling effect that was particularly disturbing when the light was reflected by bright dunes. It was therefore absolutely necessary to wear sunglasses. In the afternoon, it was impossible to tell vehicles from tanks at a distance of one or two kilometers and up. Favourable times of the day were the early morning and the afternoon hours from 1600 to 2000. In the desert, range estimation cannot be depended upon. In most cases the distances will be underestimated. It was therefore advisable to use a rangefinder.[23]

Thankfully for the Flak 8.8cm crews, rangefinders would have been integral items of kit. Yet the observation distortions must have caused significant issues for the gun crews when engaged in AT work from positions

ABOVE The pneumatic tyres of the bogies were of 812 × 165mm (32 × 6.5in) dimensions. *(Author/Axis Track Services)*

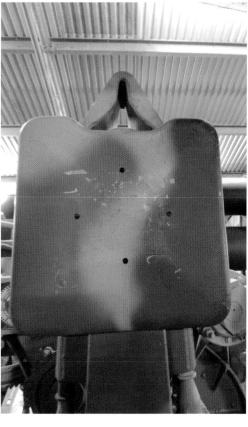

LEFT A close-up of one of the levelling jack pads. Two of the pads were of a circular design. *(Author/Axis Track Services)*

BELOW An 8.8cm gun fires near Bir Hakeim (Bir Hacheim), Libya, in June 1942, under the scorching sun. *(Bundesarchiv, Bild 101I-443-1599-20/Zwilling, Ernst A./CC-BY-SA 3.0)*

low down to the ground, the flat plane of sight across the desert surface being most susceptible to optical distortion.

While heat itself might not have been damaging to the gun, the same could not be said of probably the greatest nuisances of the desert campaign – dust and sand. At an operational level, the dust raised by the movement of the Flak's half-track transporter during daylight hours generated a dusty airborne visual marker that, for Allied reconnaissance aircraft, almost acted like an arrow pointing directly to the location of the prime mover. At least, however, the fact that the Flak guns were generally pulled by half-tracks meant that they were able to negotiate the literally shifting sands of the terrain; the crews of wheeled vehicles had to cope with the Sisyphean frustrations of continually digging their stuck vehicles out of the terrain. This did not mean that the Flak transporters had an easy life in the desert. All vehicles suffered from the effects of dust intrusion into the moving parts, clogging air filters in a fraction of the time it took in the European theatre, and curtailing the lifespan of engines by about 50–70%. (As Toppe points out, this was not just a matter of dust intrusion, but also of the fact that vehicles had to be driven long distances in low gears.) Any part of the vehicle that could be penetrated by dust, such as the brakes and chassis, had a reduced lifespan.

The gun crew had to devote much time to protecting the gun against the intrusion of

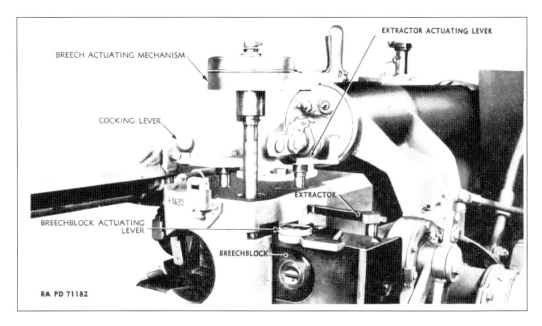

RIGHT The main operating parts of the breech mechanism. The top of the breech held the principal instruments for breech operation, cocking the firing mechanism and shell extraction. *(US War Department)*

sand and dust. Muzzle covers and plugs were installed any time the gun wasn't in action, as a build-up of sand in the gun's rifling not only affected its accuracy, it also reduced the barrel's working life and even raised the risk of serious malfunction. Most susceptible to dust, however, was the breech mechanism. A serious build-up of particles could prevent the breech-block closing, or could stop the firing pin making good contact with the shell primer. Preventing sand ingress entirely was an impossibility, but it could be limited by keeping the breech-block wrapped in a shelter-half or other piece of fabric when the gun was not in use. The daily cleaning regime had to be maintained with discipline, and all parts had to be swept clean of dust and lubricated properly. Regarding lubrication, the gunners were careful not to apply excessive amounts of oil and grease to the gun parts, as thick oily deposits became a dust attractant; once mixed with oil, dust and sand formed itself into an abrasive grinding paste that wore down metal parts over time.

Dust and sand also combined with the effects of sun glare to make the issue of visibility even more problematic. During battle, the dust thrown up by the rapid movement of troops and vehicles, the impact of artillery shells and the muzzle blast of the team's own artillery piece could transform even the brightest day into a soupy brown haze, causing issues for both anti-tank and anti-aircraft targeting. At times this could work to the Germans' advantage – disorientating the enemy as to the position of guns – but it could also drain the momentum of a gun crew's firing tempo, and it naturally had an adverse effect on accuracy and observation. For example, when firing at ground targets, the impact of the first ranging shells would kick up thick clouds of dust that were slow to clear. These interfered with making subsequent corrections on to target; effectively every shell that missed was laying down a type of geophysical smokescreen for the enemy. There was also the issue of muzzle-blast dust clouds acting as markers for the enemy, a phenomenon commented on by Toppe:

The generation of dust through recoil of the powder gases in artillery firing was of no special importance for the detection of artillery positions, because the combat zone was always enveloped in dust clouds anyway. The discharges of guns of especially flat trajectory and low-barrel elevation – anti-tank guns – could be observed because of their characteristic dust clouds. Naturally, they also prevented the gun crews from observing the effects of their own fire.[24]

This passage has an interesting relevance for the 8.8cm gun crew. For AA work and indirect artillery fire, the 8.8cm Flak would be able to limit the visual signature of their position as defined by an updraft of dust. Once they dropped their barrels into an AT mode, however, their position would become far more

identifiable, the dust kicked up in regular swirling puffs every time the gun fired. For this reason, gun crews in desert regions were constantly preparing to move at short notice once it appeared that the enemy was orienting itself to their position. In cases of AT work, therefore, it was safer to fire the gun from its wheels, and so be ready to hitch up the gun and move out after a short burst of firing.

For the Flak teams, and indeed for all troops in the theatre, the shutters would really come down in terms of visibility in the case of a sandstorm. Blowing at windspeeds of more than 100km/h (62mph), and even 150km/h (93mph) on elevated ground, the sandstorms would stir up a total 'brown out' of vision, with visibility measured in a matter of a couple of metres and even breathing becoming a challenge. On the plus side, all combat would cease during such events, but after the event the crew would need to rush into action immediately to perform an emergency clean of the weapon – both the firing position and the gun would be choked with sand, and would need rapid work to bring them back to a serviceable state.

Winter operations

Reference to winter warfare mostly looks to the extreme environment of the Eastern Front from the months of November to April, although depending on the location it was not unusual for the first snows and hard frosts to develop in October. Yet winter warfare conditions were far from just the preserve of the war out east. Flak crews would have dealt with periodic freezing combat conditions in the Balkans from 1941, Italy from 1943, northern Europe during the post-D-Day fighting of 1944–45, and during the apocalyptic fighting in eastern Europe, east Prussia and around the Baltic in the first months of 1945.

By the time the 8.8cm Flak crews reached the winter, however, they would have been through other challenges. For a start, the desert conditions that prevailed in North Africa were not totally unique to those theatres – operations during the summer months in the southern Balkans and on the Russian or Ukrainian steppe would have presented similar levels of dusty adversity to those in the Western Desert. Furthermore, the autumnal rains or

ABOVE The muzzle direction of this towed gun partially indicates that the wheeled carriage is the older *Sonderanhänger 201*. (Bundesarchiv, Bild 101I-439-1276-05/Dressler/CC-BY-SA 3.0)

BELOW January 1944. A Flak 18 sits in the freezing snows of a Berlin winter, while waiting to be towed to a new position. (Heinz Radtke/Family Archive Norbert Radtke)

ABOVE Three men on the Eastern Front lean into the task of washing out the barrel, a job to be performed after every firing. *(Bundesarchiv, Bild 101I-455-0007-23/ Kamm, Richard/CC-BY-SA 3.0)*

spring thaws brought the horrors of mud, especially challenging along the mountainous trails of central and northern Italy and the open expanses of the Soviet Union. The *rasputitsa* rainy season on the Eastern Front brought a whole new level of misery, turning all unpaved roads (bearing in mind that around 40% of Soviet villages were unpaved at this time) into wearying rivers of the thickest mud. Even fully tracked tanks, let alone the half-tracks of the Flak crews, were brought to a halt regularly by

BELOW A top view down into one of the open fuze-setting ports. The toothed pin would engage with the fuze setter on the warhead. *(Author/ Muckleburgh Collection)*

GERMAN FLAK INSTRUCTION MANUAL: KEY QUALITIES OF THE FLAK GUN CREW

The reliable and quick service of the gun is the precondition for successful action of the flak battery. When mistakes are made on the gun, then all the still exact measurements and calculations from the command post are of no use.

The first requirement that must be put on the gun cannonier is accuracy and attention. Every cannonier, even without continuous supervision, must diligently fulfil his duty at his position. He is a small wheel in the great clockwork of the battery; if the gun cannonier errs or works inexactly and inconsistently, then the action of the battery is put into question.

The second requirement requires speed and versatility of the crew with the accomplishment of their tasks. [...] A slow gun crew which cannot keep up with the loading times, makes all the calculations of the command post invalid where those certain times are used. Always, however, must accuracy go before speed. 'Quickly, however exactly.'

The third requirement that the gun cannonier cannot often enough perform, requires care of the equipment. On the gun are found many technical fine points which can only work correctly when they are properly operated, and gently and carefully handled. The cannonier who does not keep his equipment in immaculate condition makes difficulties for himself when he then has to work with that equipment. Frequent roll calls at which all the various weapons and devices of the battery received a basic inspection are an indispensable means to training the battery for the proper equipment handling.

Every cannonier on the gun must master the tasks and duties of all the cannoniers; with the loss of one, the others must each time be fully able to jump in. Locating the target is often not simple. This is not the task of the gun leader and the aiming cannonier alone, rather the task of all gun cannoniers. Every cannonier must have the ambition that his gun can be the first to report: 'Target acquired!'[27]

LEFT In the Russian winter months of late 1943, a gun crew struggle with unlimbering their gun from the front bogies. *(Bundesarchiv, Bild 101I-725-0194-21/Götz/ CC-BY-SA 3.0)*

such conditions. In terms of maintenance, gun crews would have spent much time looking after their vehicle, but also cleaning the gun frequently of mud. The combination of driving and frequent rain, plus glutinous mud, produced the perfect conditions for development of rust, so inspections were daily and protective layers of oil and grease had to be applied regularly to all moving and exposed parts.

When winter descended in earnest, however, the combination of snow, ice, hard frosts and temperatures (on the Eastern Front) of as low as −40ºC (−40ºF) could have a severe effect on the material integrity of the gun, mount and carriage components. The process of setting up the weapon had to be adapted. The chief problem in this regard related to ground frozen to the resilience of concrete. Sitting the gun mount directly on to such ground placed both gun and mount under severe strain during firing, as the unyielding soil soaked up none of the recoil. There was also the danger of the gun skidding along the hard, icy surface, hence ice spades were fitted to the outriggers to prevent slippage. It was also recommended that the earth beneath each of the outrigger feet was dug up to soften the contact between gun and ground, but in deepest winter the ground could be frozen to a considerable depth. If possible, sitting the gun on a thick bed of foliage or reeds helped to protect the gun from both the recoil and the ground temperatures.

The ingress of ice and snow into the gun was something to be avoided at all costs. As with the desert operations, in winter actions the muzzle of the gun and the breech mechanism had to be kept covered when the gun wasn't in use, and loose snow brushed off as often as was practical. Snow on the gun would melt during firing, and the water could then later refreeze to lock parts solid. It was especially crucial to keep ice out of the breech to avoid problems with loading – in the worst cases the sliding breech-block might be utterly immobile, or the firing pin might be locked in place. Under no circumstances should the gun crew have been tempted to fire the weapon with ice deposits inside the barrel, even relatively minor ones, as this could result in disasters such as blown or

BELOW This view of the front bogie of the Flak 37's limber shows the physical linkage between bogie and the gun carriage. *(Author/ Axis Track Services)*

RIGHT The marker on the recoil indicator featured three words: *Feuerpause* = cease firing; *Achtung* = attention; *Normal* = normal (recoil). *(Author/Axis Track Services)*

ABOVE The two red-capped ports on the back of the recuperator are the gas valve (left) and gas and liquid filling hole (right). The other red-capped port just visible is the drain plug. *(Author/Axis Track Services)*

LEFT The levelling jack beneath the gun carriage, raised during transit. The jacks could give 4.5 degrees of levelling either side of horizontal. *(Author/Muckleburgh Collection)*

distorted barrels or in-barrel shell detonation. The German manual on winter warfare, *Taschenbuch für den Winterkrieg*, explains that the barrel could be cleaned out using the barrel wiper tool, but for heavier ice deposits the crew might have to resort to the expedient of dissolving them with heated gun oil. It also explains that 'If firing has ceased, put the muzzle cap on the barrel to prevent it from cooling off too rapidly'; a fast transition from firing temperature to sub-zero ambient temperature could cause barrel fractures. Metal parts in general could be severely weakened by extreme cold, especially hinges and gear teeth, and therefore the crew had to discipline themselves to free frozen components by defrosting rather than force, even though battle fatigue and sleep deprivation must have made simply whacking stuck parts with a mallet wearily tempting.

Once the thermometer readings dropped below freezing, the Flak crews also had to implement special measures regarding the gun's fluid content, specifically in the recoil and recuperator mechanisms. (General procedures regarding the care and maintenance of the Flak 8.8cm guns are explained in more detail in Chapter 7.) If the liquids inside these components were frozen, the disastrous outcomes ranged from a broken gun mount to cracked outriggers (other components would be forced to absorb the gun's recoil), so the fluid composition had to be adjusted to resist freezing. The *Taschenbuch für den Winterkrieg* makes some general points applicable to the Flak 8.8cm as well as other weapon types:

The brown recoil liquid can be used for nonautomatic guns in temperatures as low as –40 degrees F., but for pieces with semiautomatic breech mechanisms, such as the 50-mm antitank gun 38 (5-cm Pak 38), only as low as –4 degrees F. At lower temperatures, the recoil brake and the compressed-air counterrecoil mechanism should be filled with cold-resistant recoil liquid in order to guarantee the opening of the semiautomatic breech mechanism and the expulsion of the cartridge case after the counterrecoil. The filling of the recoil brake and counterrecoil mechanism should be done only by the armorer or assistant armorer. Antifreezing solution, when used in guns, is cold-resistant down to −67 degrees F. For guns whose recoil brakes are required by regulations to be filled with brake oil, Shell oil AB 11, which is frost-resistant down to −76 F., should be used. This oil can be mixed with captured oil without special precautionary measures.[25]

In addition to adjusting the fluid mix for the recoil and recuperator systems, the Flak crews and engineers had to change the composition of their general oils and lubricants. Unmodified, the standard German gun-cleaning oil could be used down to temperatures of −30°C (−22°F) while for lubricating oil it was −20°C (−4°F). For even lower temperatures, the Germans adopted the expedient (long used by the Soviets) of adding kerosene to the oil in the proportion of 2:1; this simple additive ensured that the oils still functioned properly in temperatures as low as −40°C (−40°F). Kerosene could also be added to lubricating grease for similar performance enhancements. In relation to finer grades of oil, the winter manual explains that:

Oil for delicate mechanisms, such as the aiming mechanism, is frost-resistant down to −40 degrees F. The troops in the east and north are issued for winter use Vacuum Servöl 222, *which is frost-resistant down to −58 degrees F. In case of emergency, gun-cleaning oil may be used. Complete greasing of the aiming mechanism, preceded by the removal of the old oil, may be done only by the armorer or the assistant armorer.*[26]

In addition to paying special care to the maintenance of the Flak 8.8cm, the gun crew also had to keep a close watch over the condition of the ammunition they were feeding into the gun. It was all too easy to miss the build-up of clear ice on the outer body of a shell, ice that would either prevent the shell from being rammed properly into the chamber

ABOVE A colour-coded version of the *Übertragungsgerät 37* elevation transmission receiver. *(Author/Axis Track Services)*

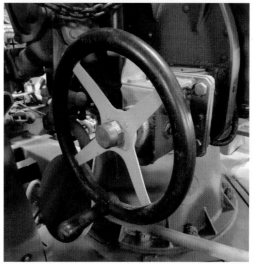

LEFT The traverse assembly, with the high gear selected, could turn the gun 360 degrees in 33.9 seconds. *(Author/Axis Track Services)*

or would affect its accuracy in flight. It was especially important to remove any ice build-up around the driving bands, as this would compromise the seating of the shell and the proper engagement of the shell with the rifling.

Above we have touched on just some of the theatre challenges that faced Flak 8.8cm crews. Admittedly we have scarcely engaged with the myriad human factors that placed additional burdens upon the shoulders of already weary crews – such as hunger, extreme fatigue, homesickness, combat stress, fear, interpersonal issues, and so on. Some of these issues will emerge in the following two chapters, where we explore the 8.8cm Flak's operational role in the Second World War, taking a closer look at some of the battles the guns fought both at the micro and the macro levels.

Chapter Five

At war – anti-aircraft operations

In every theatre in which they served, the 8.8cm Flak guns made the skies a hostile place for Allied aircraft. As part of a large integrated air defence system, they could virtually carpet the sky with exploding ordnance.

OPPOSITE Flak batteries were positioned around the industrial areas most likely to be hit by Allied bomber attacks. Note the AA gun mounted on a rail carriage, which usefully allowed for gun repositioning. *(AirSeaLand Photos/Cody Images)*

ABOVE A Flak 37 in a tidily prepared anti-aircraft position somewhere in Berlin. Berlin's public spaces were littered with AA guns by mid-1943. *(Heinz Radtke/Family Archive Norbert Radtke)*

The 8.8cm Flak guns were not standalone weapons in terms of Germany's air defence, but were part of a graded system of gunnery, which can be categorised according to light and heavy types. Light Flak weapons were typically heavy machine guns or 2cm cannon, such as the 2cm Flak 30/38, and a range of 3.7cm guns, including the 3.7cm Flak 18/36/37 and Flak 43. These guns were perfect tools for engaging low-flying and fast aircraft out to altitudes/ranges of approximately 500m–5,000m (1,640–16,404ft). As automatic weapons they could put up a tremendous volume of shot – some had a practical rate of

RIGHT The intended result. A USAAF B-26 Marauder of the 9th Bombardment Division goes down in flames following a Flak strike. *(AirSeaLand Photos/ Cody Images)*

fire of up to 180 rounds per minute – using volume of fire to search out their often blurring, evasive targets.

The next stratum up was, in fact, the medium AA weapons, into which category we mainly place 5cm guns, such as the 5cm Flak 41. Such guns were developed to straddle the perceived gap between the capabilities of the light AA guns and the heavy guns at the top, with a maximum ceiling of up to 9,000m (29,527ft). Yet the medium AA gun was, in reality, a failed area of the development for the Germans, with the weapons produced being of unsuccessful designs or of very limited production numbers, or both. So it was that the lion's share of the AA work was performed by the light guns and also the heavy guns. 'Heavy' AA guns referred to the 8.8cm Flak, the 10.5cm pieces (10.5cm Flak 38 and 39) and the 12.8cm Flak 40. As single-shot weapons, the heavy guns could not individually lay down the volume of fire of the light AA guns, but they could reach up to progressively high altitudes: the Flak 18/36/37 to about 8,000m (26,246ft), the Flak 41 and the 12.8cm Flak 40 to 10,675m (35,022ft) and the 10.5cm Flak 38 and 39 to 9,450m (31,003ft).

Organisation of Flak units

The 8.8cm gun was the most prolific German heavy AA gun during the Second World War. Giving this some context, we should also acknowledge the sheer importance and extent of the Flak arm during the Second World War, pushing against the tendency to sideline AA fire in favour of focusing on the more dramatically stirring air war. The Flak branch of the Luftwaffe, at its peak, numbered about 1.25 million personnel, which constituted about half of the Luftwaffe's total manpower. The formations and units of the Flak served in every theatre in every imaginable terrain, combat environment and level of danger, from manning relatively quiet (at least until June 1944) French coastal batteries to fighting near-suicidal actions on the Eastern Front. Yet even disregarding the 8.8cm's extensive applications as an AT weapon (the subject of Chapter 6), its role as an AA weapon was utterly central to the defence of the Reich and the occupied territories, and especially against the swelling thunder of the Allied strategic bombing campaign of 1943–45. Indeed, military historian

LEFT This artwork gives a sense of the strangeness of night-time AA work, with the landscape lit up by a strobing flash with each firing of the Flak guns. *(AirSeaLand Photos/Cody Images)*

ABOVE **Having received an alert for an approaching air raid, an 8.8cm Flak gun crew race to their positions.** *(AirSeaLand Photos/Cody Images)*

Steven J. Zaloga has called the 88 gun 'the basis for national air defence'.

Before looking at the types of AA fire that an 8.8cm battery could deliver, we need a sense of how the Flak arm was rationalised in general. Underpinning the service structurally was the fact that Flak served both in a mobile context, moving with the front-line field armies, and in static defence – that is, sited semi-permanently around towns, cities, critical industrial facilities (especially oil refineries and ports), coastal positions and any other locations of significance (V-1 and V-2 rocket sites, for example). The Luftwaffe had by far the largest portion of the Flak pie; the Kriegsmarine (Navy) and the Heer (Army) both created their own organic Flak units, but the former was largely consigned to port protection while the latter tended to be small tactical units with divisions. Note, however, that the Heer could and did have operational claim over Luftwaffe field units, and from 1944 its authority to requisition Flak units for front-line ground combat use could trump the

RIGHT **The 8.8cm Flak guns were typically supported by fast-firing 2cm Flak cannon, which dealt with low-flying aircraft such as fighters.** *(AirSeaLand Photos/ Cody Images)*

needs of the Luftwaffe home defence (see below and Chapter 6).

In terms of the formation/unit organisation for Luftwaffe Flak, the top of the hierarchy was occupied by the *Korps* headquarters overseeing two to four *Divisionen*. In front-line service, the Flak divisions would typically be assigned to serve as part of a *Luftflotte* (Air Fleet), but often seconded to work within an Army corps or similar large formation. The relationship was a complex one, as a 1943 US military intelligence document acknowledged:

> In the field, antiaircraft artillery is operationally subordinate to the commander of the army to which it is attached, while remaining subordinate to the German Air Force for administration. Its use in cooperation with the army is extremely flexible, the scale and method of employment being varied, frequently at very short notice, according to the tactical situation. In general an antiaircraft artillery corps works with an Army Group, the chain of command being exercised through

ABOVE The cause of this barrel explosion in France in June 1944 is unknown (at least by the author). If it exploded at this angle, then a premature shell detonation or barrel malfunction would be the likeliest explanation. *(AirSeaLand Photos/ Cody Images))*

LEFT Two 8.8cm guns wait for a visit from RAF Bomber Command, their muzzles angled skywards, as night settles over an industrial target area. *(AirSeaLand Photos/ Cody Images)*

RIGHT A 150cm searchlight sends out its powerful beam. There would be close cooperation between Flak and searchlight teams, the latter attempting to 'cone' targets for the former. *(AirSeaLand Photos/ Cody Images)*

antiaircraft artillery divisional and regimental staffs down to the battalions. Although no hard-and-fast rule can be laid down, an antiaircraft artillery division generally works with an army, and a regiment with an army corps; individual battalions are allotted to army divisions, preference usually being given to armored and motorized units.[28]

As we shall see later in this chapter, and in the next, the fact that the Flak arm could be partially absorbed into the Army tactical requirements

RIGHT Sound location equipment may look crude to modern eyes, but used properly in good weather conditions it could provide range-bearing information for targets out to 6km (3.7 miles). *(AirSeaLand Photos/ Cody Images)*

LEFT Searchlights and acoustic locators worked closely together, the latter providing a stream of information to assist the accuracy of the former. *(AirSeaLand Photos/Cody Images)*

brought considerable cost for the AA gunners. For home defence (AA) service, meanwhile, the divisions usually served a *Luftgau* headquarters. The *Luftgau* was a regional command, and hence the Flak arm's weapons could be divided up logically to protect specific geographical zones and to interdict predicted and established enemy flight paths.

Within the individual *Flakkorps*, the 8.8cm guns and other weapons were arranged by two(+) *Brigaden* and the brigades by two to four *Regimenter*. Regiments had four to six

BELOW A 150cm searchlight team in position. The searchlights could detect targets out to 8km (5 miles), depending on the weather conditions. *(AirSeaLand Photos/ Cody Images)*

ABOVE The basic Flak battery consisted of four guns arranged in a square pattern, as seen in this wartime artwork. *(AirSeaLand Photos/Cody Images)*

RIGHT The relationship between the gun and the aircraft here is uncertain; it is likely to be purely a bloodless aiming and firing drill. *(AirSeaLand Photos/ Cody Images)*

Abeteilungen (battalions), depending on whether they were static or motorised (the latter had four battalions). Battalions were classed as either *Schwere* (heavy), *Leichte* (light), *Gemischte* (mixed – both heavy and light guns present, typical in front-line units) – and *Scheinwerfer* (searchlight battalions). The 88s belonged to either the *Schwere* or *Gemischte* battalions. Within a heavy battalion, the guns were further organised into batteries of, typically, four to six guns each.

The broad brushstrokes of German air defence policy are beyond the remit of this study (the Bibliography at the end of this book provides further reading on this subject). What is relevant to explain here is how the 8.8cm batteries structured themselves for the AA war, and how the doctrine broadly evolved as the threat from the air grew in magnitude for the Reich. At the beginning of the war, the basic home defence AA battery structure consisted of four guns arranged in a square pattern with the command post (which included the predictor) placed in the centre. By 1941 the command post had shifted about 100m (328ft) away from the guns, where its personnel could concentrate on their tasks and man their sensitive equipment away from the noise and muzzle blast of the guns. What we see between 1942 and 1945 is a steady expansion in the size options of the battery, as the weight of Flak fire had to be scaled up to meet the sheer numbers of Allied bombers. Six-, eight- and even twelve-gun batteries were developed – the latter known as *Grossbatterien* – these controlled by multiple predictors and one or two radar units and arranged in various rectangular or circular patterns, depending on the need. (Radar became a major presence in Flak gun control by the spring of 1942, although lack of sufficient radar for each battery was one reason behind the development of the *Grossbatterien*.) The batteries would be arranged in a variable pattern around a central point deemed particularly vulnerable, such as a factory complex. Specifically, the guns would orientate themselves to engage the enemy bomber during the roughly 6km (3.7 miles) of flight up to the point of bomb release, the aim being if not to destroy the enemy aircraft, then at least severely disrupt their flight path and thus the accuracy of their bombing.

LEFT A Flak 18 loader prepares to ram a shell into the breech, with the gun at full elevation. Personnel would rotate through the loading role during intense firing periods. *(AirSeaLand Photos/ Cody Images)*

BELOW January 1940 – a new Luftwaffe AA crew take their oath of service, with a Flak 18 and a 2cm Flak 30 in the background. *(AirSeaLand Photos/ Cody Images)*

Controlling the fire

While engaging the enemy bomber force, the focus of the 88s was to deliver as many time-fuzed HE shells as possible into the midst of the bomber stream, the shells detonating at the correct altitude and in or along the enemy flight path to maximise the possibility of a hit or destructive near-burst. Shells bursting within 9m (29.5ft) of the enemy bomber would likely result in the death of the aircraft, through a combined fragmentation and blast effect, but fragmentation damage to aircraft could occur at distances two or three times greater, tearing ragged holes in airframes, crippling engines and damaging flight surfaces.

Accurate fire control was central to the combat results of the 8.8cm Flak guns, as AA gunnery was all about the business of prediction. Roughly speaking (and averaging out), an 8.8cm shell would take 1 second to climb 305m (1,000ft), thus against bombers flying at 7,620m (25,000ft) there could be a 25-second flight interval between the gun firing and the shell reaching the target area. During this interval, the bomber would keep flying on at speeds of around 400km/h (250mph), meaning that it would cover nearly 3.2km (2 miles)

ABOVE A Flak 18 emplaced in a country lane in Belgium. Note the rock placed under the outrigger jack plate to help with levelling. *(AirSeaLand Photos/Cody Images)*

RIGHT Houses in the immediate vicinity of an AA battery would be prone to structural damage from the continual blast of the guns. *(AirSeaLand Photos/Cody Images)*

LEFT This excellent pre-war photograph shows a Flak 18 at Nuremberg conducting a demonstration AA drill. Note the large number of wicker shell containers. *(AirSeaLand Photos/Cody Images)*

BELOW An 8.8cm Flak oriented for anti-aircraft fire. To the right are what appear to be a 5cm Pak 38 and a 3.7cm Pak 36. *(AirSeaLand Photos/Cody Images)*

of distance by the time the shell reached its altitude.

Here was the essence of the cat-and-mouse game between the 8.8cm Flak teams and the high-flying bomber. The basic flow of fire-control information would first involve picking up the bomber aircraft either visually or by radar, and then tracking them by these means while the rangefinder or radar provided altitude information. All this data would then flow into the director system, which would feed the correct elevation and traverse information, plus altitude details for the fuze setters, through to the gunner's receivers.

It was always recognised that however accurate the calculations received from the director might be, high-altitude fire would always be an imprecise business. At altitudes of 7,620–10,668m (25,000–35,000ft), errors in barrel orientation measured in mere centimetres would result in misses by many metres at the target height. Furthermore, the Allied bomber

ABOVE This excellent photograph, again taken somewhere in Berlin, shows a Flak director team with the sizeable *Kommandogerät 36* (KDO. GR. 36; Stereoscopic Director 36). *(Heinz Radtke/Family Archive Norbert Radtke)*

ABOVE RIGHT In an effort to catch the drama for propaganda purposes, this German cameraman trains his telephoto lens on the same elevation as the neighbouring 88 gun. *(Bundesarchiv, Bild 183-2008-0415-500/CC-BY-SA 3.0)*

crews, wise to the patterns of AA fire, would make alterations in altitude, heading and speed at regular prescribed intervals, thwarting the efforts of prediction below. Kill rates were consequently very low when one considers the volume of fire – about 2,500–3,000 8.8cm shells were fired for every kill achieved.

To compensate for the inaccuracy, the Flak batteries had to put as high a volume of flak as possible into the bombers' general flight path, and to adjust the fire constantly so that the explosions tracked the aircraft – one battery (or, more accurately, one group of batteries) passing over to the next when the aircraft began to pass out of their effective range envelope. If done well, 'continuously pointed fire' (as the British and Americans called it) subjected the bomber crews to a hellish experience, with Flak exploding constantly around them and getting ever more accurate as the Germans fine-tuned the fire on to their target. The worst time would be during the bomb run, which required straight and level flight, presenting the optimal moment for the German gunners to kill the aircraft or at least disrupt the bomb run.

There were some other fire patterns that the

RIGHT Flak guns would be positioned along every avenue of approach around important industrial plants. Some major installations, such as oil refineries, would have literally hundreds of artillery pieces devoted to their protection. *(AirSeaLand Photos/Cody Images)*

LEFT In this rare photograph, taken on a summer's day and labelled 'Western Front, 1940', a frame has been fitted over the barrel of a Flak 18 to attach camouflage. *(AirSeaLand Photos/Cody Images)*

BELOW This 8.8cm Flak 18, in France in 1940, has nearly the full complement of crew in the picture – ten men in total. The photographer was probably also a crew member. *(AirSeaLand Photos/Cody Images)*

German AA teams could apply. One of these, predicted concentration, involved several batteries linked by a common fire-control centre. Instead of firing at the bomber throughout the duration of its flight, the guns would all be orientated to engage a certain point in space where the bombers were expected to be at a given moment in time. Once the guns had fired their first shells at this point, they would then be adjusted to aim at a new coordinated point, and so on. For the enemy bomber crew, the experience would be a sudden and heavy ripple of multiple simultaneous Flak explosions at regular (about 60 seconds) and unnerving intervals, with relatively silent gaps in between, notwithstanding the onslaughts of enemy fighters.

Prediction was still a problem with any form of predicted-concentration fire, especially with the enemy aircraft making evasive manoeuvres. Thus the Flak teams could opt for a different type of fire, again known by the Allies as

ABOVE An air-raid alarm in Greece sends the crew of a *Kommandogerät 36* sprinting from their crude tents to the director. *(AirSeaLand Photos/Cody Images)*

about 2km (1.2 miles) away, working closely together through radio and communicating constantly with radar personnel. The ideal was for one of the searchlights to detect and 'lock on' to an enemy bomber, then the others would also highlight the target to 'cone' it for the AA gunners to fire upon visually. The searchlights also had a secondary purpose of obscuring the vision of the bomber pilot, navigator and bombardier, making their maintenance of an accurate bomb run difficult if not impossible.

As any Allied bomber pilot would attest, dealing with Flak was one of the most harrowing experiences. Although the evasive-action manoeuvres could and did make a difference, sometimes the sheer weight of Flak fire – which might seem to fill almost every available gap in the sky during a daylight raid – made it feel more a matter of luck than judgement whether the bomber survived the hail of shells. For a personal insight into what it was like to be on the receiving end, B-17 airman Lloyd Krueger recounts his experience of an air raid over Czechoslovakia on 12 May 1944:

'barrage'. Here the fire-direction team accepted a certain degree of inaccuracy in their shooting, and to compensate they would designate a certain three-dimensional 'box' of space in the enemy flight path, into which the Flak guns would deliver a dense rolling concentration of fire. (A box shape is actually just one example; the German gunners could also lay down barrage fire as screens, cones or other shapes.) As the barrage method was the less accurate of the two, it would usually be concentrated into the space immediately in front of the anticipated enemy bomb-release point, the most predictable part of the flight path.

The delivery of Flak fire naturally changed between day firing and night firing. At night, the role of the searchlight battalions was critical. The searchlights were usually arranged in groups of three in the pattern of a triangle, consisting of a master light at the apex of the triangle and the other two at the lower points

At 1134 hr. the 95th reached the I.P. Point near the town of Chomutov, Czechoslovakia. We turned on a heading of 52° for our eight-minute run to the target. We were immediately greeted with a wall of flak. Each burst resembled an inverted 'Y' that was created in fuzzy black smoke and that appeared to be about ten foot high, with all legs of this figure about a foot wide. Usually, if you saw a single burst, it was soon to be followed by four or five additional ones. I suspect that they inserted clips into their gun with four or five 88mm shells. By the time we could see these ominous shapes of smoke, the shell had already burst into thousands of tiny pieces of shrapnel, each spinning into space and searching for something to pierce or tear apart. Sometimes larger chunks would be thrown out in our direction. Anyone of these pieces could easily pass through the thin aluminum skin of our B-17 and still have the force to penetrate the body or some vital part of the plane itself. The flak we were facing today was a new experience and one that we all learned to live with. We could not fight back at it nor in any way stop it.

LEFT Flak towers, such as this one at Augarten in Vienna, were monstrous bomb-proof structures with AA gun platforms on the roofs. Although 8.8cm guns were occasionally sited on top, heavier 10.5 and 12.8cm weapons were more common. *(Luftbildfotografie FujiUser)*

BELOW The proximity of AA guns to urban areas meant that during air raids tons of Flak shrapnel would fall back down on to the city. *(AirSeaLand Photos/Cody Images)*

ABOVE A demonstration of AA weaponry to German factory workers in 1940. The three main pieces of equipment here are a Flak 36/37, a 150cm *Flakscheinwerfer* (Flak searchlight) 34 and an RRH acoustic locator. *(AirSeaLand Photos/Cody Images)*

The only challenge we could extend to the gunners of the enemy was to throw out chaff. Chaff was small pieces of aluminum foil that was cut up and bundled. Our gunners would throw this out of the B-17 and let it drift slowly earthward. This chaff could be picked up on the German radar that determined the altitude of the bomber formations and that set the timing of when these 88mm shells would explode. Chaff sometimes would lower the bursts, therefore, helping the high squadrons and perhaps hurting the lower ones. Once you turn into the target from the I.P. Point, the formation cannot carry out evasive action or deviate whatsoever from the course. It is during this straight run that the lead bombardier is getting the target set in his Norden Bomb Sight and is determining just when to release these eggs of destruction. The German gunners knew that these formations of planes would be on a constant heading, so it was not hard for them to place walls of flak along the entire length of our bomb run.[29]

Nor was it just the bombers that were vulnerable to the 8.8cm Flak guns. In the following account, by P-51 Mustang ace Eric Hammel, US fighters are engaged abruptly while flying over France in August 1944:

We flew a little way into Germany, turned northwest for a while, and set course for A-2. On the way home I could see Reims off in the distance to our left. Nobody said anything about it as we passed. We were on a course that was taking us about twenty miles north and west of Paris. We had been airborne about two and a half hours. No one had said anything on the radio. We had not received any flak, and we had not seen any targets of opportunity.

As we crossed the Seine River at about 12,000ft, four rounds of 88mm flak were fired at us. One round hit Wally Emmer's brand-new D-model Mustang square in the fuselage tank, just behind the cockpit. His airplane exploded into a great fireball and the flaming debris flew in all directions. The right wingtip passed across my airplane, between the prop and the windshield.[30]

This account reminds us that although the 88s relied to a certain degree on luck when firing at high-altitude targets, they were still highly accurate weapons that could hit aircraft with a high degree of precision at the lower altitudes.

ABOVE A battery of Flak 18s open up on British bombers during the fighting in France in the spring of 1940. By this time the 8.8cm Flak was also building its reputation as a tank-killer. *(AirSeaLand Photos/Cody Images)*

LEFT The *Kommandogerät 36* was a very personnel-intensive piece of equipment. Here 11 people are actively engaged with gathering target data in Greece in 1940. Note the additional stereoscopic rangefinder in the background. *(AirSeaLand Photos/Cody Images)*

RIGHT The range drum mechanism of the *Kommandogerät 36* director, which provided horizontal range and elevation calculations. *(AirSeaLand Photos/Cody Images)*

RIGHT The target course plate was used to keep continual track of enemy aircraft's bearing and speed. *(AirSeaLand Photos/Cody Images)*

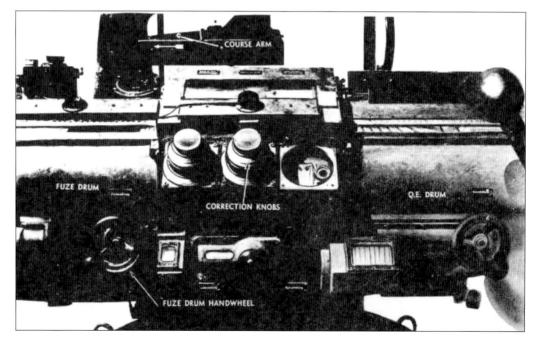

RIGHT A US manual diagram of the wind correction knobs on the *Kommandogerät 36* director. *(AirSeaLand Photos/Cody Images)*

ABOVE A Flak 18 in one of the German Westwall defensive positions in August 1939, shortly before the German invasion of Poland. *(AirSeaLand Photos/Cody Images)*

The effectiveness of the 8.8cm Flak

It is impossible to say exactly how many Allied aircraft were downed specifically by 8.8cm guns, as opposed to other types of AA weapon. Yet we can make some educated generalisations about their performance, particularly in relation to their deployment against the Anglo-American strategic bombing campaign. We start by noting that the 88s constituted the largest part of the Reich air defences. In August 1944, there were 13,260 heavy AA guns in Germany, and a total of 10,930 were 8.8cm Flak guns.[31] Although this figure is taken from one particular point in time, we can reasonably safely assume that, based on production figures of heavy AA types, that the 82% composition of 88s was relatively representative, and would likely have been even higher in the early years of the war.

Looking at the losses for the strategic bombing campaign in total, we can make a clear distinction between the night-bombing campaign of the British and the day-bombing campaign of the Americans. Taking the former first, during its night-hours bombing campaign RAF Bomber Command lost 3,623 aircraft (June 1942–April 1945).[32] Of these, 1,345, or 37%, were accounted for by Flak; the rest were shot down by German night-fighters. Now if we look at the US figures, and their day-bombing strategy, a signal difference emerges. In the ETO, Flak shot down around 5,400 US aircraft, while fewer – 4,300 – were accounted for by the fighters. Looking at some of the specific air formations, the 8th Air Force had 1,798 aircraft shot down by AA fire, while

RIGHT This acoustic locator consisted of four sound-gathering horns – two vertical and two horizontal – with a technician assigned to each pair, one either side. The stereo inputs enabled the technicians to make elevation and bearing calculations. *(AirSeaLand Photos/ Cody Images)*

BELOW A *Kommandogerät 36* in position on the island of Elba, Italy. The top of the director is dominated by the *Raumbildentfernungs- messer (Höhe)* – Em. 4m. R (H) – the 4m stereoscopic rangefinder. *(AirSeaLand Photos/ Cody Images)*

the Mediterranean Allied Air Forces (MAAF) lost 2,607. Westermann explains in detail that 'Specifically, the Fifteenth Air Force lost 1,046 heavy bombers to flak between its activation in November 1943 and its final bombing mission in May 1945. The heavy bombers lost to flak represented 44 percent of all Fifteenth Air Force heavy bomber losses. Approximately 10 percent of these losses occurred during attacks on the oil vicinities in the vicinity of Ploesti, the "graveyard of bombers," the vast majority as a result of flak. In addition to the strategic air forces' loss of heavy bombers, Luftwaffe flak claimed a total of 2,415 aircraft from the Ninth and the Twelfth Air Force'[33]

Taking these figures and aligning them with the fact that the bulk of the heavy AA consisted of 8.8cm guns, then we can see that the specific Flak weapons studied in this book were indeed the bedrock of German air defence, their relative inaccuracy over great range compensated for by their weight of fire, and the losses they inflicted were comparable to or even exceeded those inflicted by the German fighters. The figures above show that the day bombers were more at risk from Flak than the night bombers, for obvious reasons – it's always easier to hit something if you can see it. Furthermore, if we include the numbers of aircraft the Flak guns damaged, then the picture of their influence becomes even more

ABOVE Full camouflage paint is rarely seen applied to a Flak gun, but this Flak 18 – emplaced in a permanent AA position in Germany – is an exception. *(AirSeaLand Photos/Cody Images)*

profound. In total, the US 8th and 15th Air Forces suffered 66,493 aircraft damaged by Flak. Although the level of damage could vary from minor airframe holes to major structural disarray, the fact remains that the Flak effects must have consumed tens of thousands of ground personnel man hours, as well as putting hundreds of aircraft out of commission at any one time. The 'damage' inflicted on the Allies by the exploding Flak would have also included thousands of aircrew, dead and wounded.

In the final analysis, however, the 8.8cm Flak resources of the Third Reich were simply not sufficient to handle the demands of the war. The guns and crews dedicated to homeland air defence were progressively drawn away into field service, especially as the Soviet juggernaut came closer and closer to the German homeland. Westermann notes that 'the Luftwaffe transferred a total of 555 heavy and 175 medium/light batteries to the fighting fronts during the last eight months of the war. The mass transfer of flak batteries to the combat front in the closing stages of the war effectively stripped entire areas within Germany of their air defenses, opening these areas to unimpeded aerial attacks.'[34] Ultimately, it was the very capability of the 88 to slip between ground-fighting and air defence roles that proved the final undoing of so many of its crews.

LEFT Two gunners do some manual fuze setting, adjusting the timing mechanism duration and preparing the shells for firing, December 1940. *(AirSeaLand Photos/Cody Images)*

Chapter Six

At war – anti-tank operations

Designed as an AA weapon, the 8.8cm Flak guns nevertheless forged their notoriety in ground warfare. While there was no doubt about the armour-killing power of the '88', reality and mythology must be prised apart before we can arrive at a proper appraisal of the gun's significance as an AT weapon.

OPPOSITE The 'kill rings' around the barrel of this Flak 18 indicate that it saw some heavy fighting before its capture by US forces in Germany, 1945. Note the shells sitting in the fuse setters. *(AirSeaLand Photos/Cody Images)*

ABOVE A Flak 18 performs a ground bombardment during the invasion of Poland in September 1939. The gun is at full recoil. *(AirSeaLand Photos/Cody Images)*

Although it was in North Africa from 1941 that the Flak 8.8cm guns really became notorious as AT weapons, they had already been tested in this combat role during actions in the Spanish Civil War (to a limited extent), Poland in 1939 and in the western Europe campaign of 1940. The German *Blitzkrieg* in Poland, launched on 1 September 1939, saw the infantry, armour and air units supported by 2,628 AA guns, including a large number of 8.8cm Flak guns. Although history books can present the German invasion of Poland as an easy victory for the Wehrmacht, there was in fact extremely hard fighting that forced the Germans to adapt both equipment and tactics. Luftwaffe Flak batteries found an immediate utility in providing direct fire against Polish infantry and artillery positions, and also in beating off some of the attacks from light Polish armour. One of the landmark actions occurred around the village of Piłatka on 8–9 September, when the German 3rd Light Division, pushing forward in an aggressive reconnaissance action, came under attack from massed Polish infantry and artillery, forcing the Germans into defensive positions. The 8.8cm guns of Flak Regiment 22 were deployed to bolster the defence, the guns and their teams spreading out around positions south of the Piłatka – Iłża round. There they came under persistent attack by both day and night, but the Flak 18 guns – set up in staggered positions with good fields of fire – were soon wrecking Polish armour and suppressing infantry attacks. Make no mistake, the gun crews were in the thick of the fighting, and the assaults closed to

RIGHT A view of the recoil indicator on the 8.8cm Flak at the Muckleburgh Collection. The gun's firing level is on the left-hand side. *(Author/Muckleburgh Collection)*

such range that the Luftwaffe gunners at times fought the Polish infantry on equal terms, with small arms, bayonets and grenades. Under the relentless pressure, however, the Flak guns were eventually withdrawn, three of them having to be left behind.

The combat around Piłatka was one of several ground actions fought by the 8.8cm Flak guns in Poland. The weapons were even used in the indirect-fire bombardment role during the pummelling of Warsaw in the final stages of the Poland campaign. Although it does not appear that the Luftwaffe or Wehrmacht consciously rebranded the purpose of the 8.8cm guns to an AT weapon on the basis of these actions, it was clear that the Flak guns were tactical chameleons, shifting roles dependent on context and requirement. As if in implicit acknowledgement of this fact, 8.8cm Flak guns began to be fitted with protective armour shields in late 1939 and early 1940. Being frontal shields, this additional protection spoke of the fact that the gun crews might now find themselves within small-arms range, which meant that they were engaging ground targets rather than air targets (enemy aircraft typically attacked German rear-area logistics, where the Flak guns were safely out of the range of small-arms weaponry).

The profile of the 8.8cm Flak as a 'general-purpose' artillery piece was reinforced, and its reputation enhanced, during the German campaigns in western Europe in 1940, where the 8.8cm guns confronted British and French armour on a significant scale. During the initial phase of the German invasion of France and Belgium, the 88s were applied for the purpose of bunker-busting, destroying French fortified positions along the Meuse river when infantry attacks and dive-bombing from Ju 87 Stukas had proved ineffective. The great virtue of the 8.8cm Flak in this role was its armour-piercing ammunition, which proved comfortable in handling ferro-concrete positions, plus its flat trajectory and accuracy. The contribution of the 88s to this moment of the campaign was locally decisive. Around Floing, a village west of Sedan, a single 88 was manhandled up to a firing position near the river, and proceeded to destroy French bunkers that had up to this point shrugged off the fire of Panzers and tank-destroyers, most of which were armed with 7.5cm guns. The contribution of the 88s to the Meuse crossing was critical, and it ensured that the German casualties during this vulnerable phase of the operation were kept to a minimum.

It was at this point that the British in particular began to acquire what became an inveterate anxiety about the 88s, and the Flak guns embraced the anti-armour role with growing confidence. One of the landmark moments for the 88 was its role in stopping the Anglo-French counter-attack launched on 21 May around the town of Arras. Although the attack was flawed from the outset in force composition and tactics, it still posed a major threat to the German advance, the British throwing in the weight of more than 70 tanks – Matilda Is and IIs – plus some French SOMUA S35s. The standard German AT gun at the time, the 3.7cm Pak 35/36, soon proved inadequate to the task of penetrating the British armour, so a certain Generalmajor Erwin Rommel, commanding the 7th Panzer Division, brought up his divisional 8.8cm Flak support guns, plus those of the Flak Regiment 22, and set up a staggered gunline, coordinated through wireless communications. It was these guns

BELOW A French tank 'brews up' following strikes by 8.8cm shells in an action at near point-blank range. The gun crew are using one of the bogies as improvised cover.
(AirSeaLand Photos/ Cody Images)

RIGHT France, 1940. Following the successful German occupation of France, a Flak 18 is manoeuvred into a defensive position over Marseille harbour. *(AirSeaLand Photos/Cody Images)*

more than anything else that wrecked the Allied armoured drive, the 8.8cm AP shells having no problem punching through even the Matilda II's 78mm (3in) frontal armour. In just a few minutes of the 88s opening up, 24 British tanks had been destroyed. This fire, plus the relentless contributions of the Luftwaffe, finally broke up the Allied ambitions, and forced the surviving armour into retreat. Significantly, the 8.8cm gun was also now implanted as an AT gun in the mind of Rommel and his superiors.

North Africa and the Western Desert

As an AT weapon, the 8.8cm Flak guns had some genuine limitations. Being 'heavy' artillery pieces, they were most suited for fighting from emplaced positions from their cruciform platform or from the fixed /2 mount. An experienced crew could transition the weapon to and from its ground mount in a matter of minutes, but in a fluid combat

BELOW An 8.8cm Flak 18 in the North African desert. The men on the right are preparing ammunition for use, fitting fuzes. *(Bundesarchiv, Bild 101I-443-1574-23/Zwilling, Ernst A./CC-BY-SA 3.0)*

LEFT The spent shell cases speak of the intensity of fighting around this captured German Flak position in North Africa. *(AirSeaLand Photos/ Cody Images)*

action with enemy armour bearing down on the gun position, those minutes could make a mortal difference for the gun crew. The gun could also be fired from its wheels, but this placed a severe strain on the carriage, and also comprised the gun's accuracy. Furthermore, the 8.8cm Flak guns were not low-profile weapons; they presented a high silhouette, made higher when seated on the wheels. Digging the gun in and the judicious use of camouflage would help break up the profile of the weapon to the enemy, but once the firing started the opposing infantry, armour and artillery would typically have little problem in identifying the Flak positions and orienting their fire towards them.

For these reasons, the 88 excelled in unbroken and flat terrain where it could use its long range and flat trajectory to engage the enemy well before it reached close quarters. Hence the 8.8cm Flak was perfectly at home in the fighting in Libya and Egypt. Although climatic and environmental conditions in the

BELOW With the front bogie already removed, the gun team manhandle the rear bogie out from under the emplaced Flak 18. *(Bundesarchiv, Bild 101I-724-0135-13/ Briecke/CC-BY-SA 3.0)*

ABOVE An abandoned Flak 18 in the North African desert, with a single 8.8cm armour-piercing shell that never made it into the breech. *(AirSeaLand Photos/Cody Images)*

entitled 'A Tactical Study of the Effectiveness of the German 88 mm Anti-Aircraft Gun as an Anti-Tank Weapon in the Libyan Battle'. The report was basically an analysis of how the British had been faring against the weapon. US observers nervously studying the fighting around Tobruk gave some sobering conclusions, one commenting: 'The German 88mm guns penetrate the armour of all British tanks. British tanks dare not attack them. Up to now the British seem incapable of dealing with these weapons.' Another observer provided a more detailed account of an action:

Western Desert meant that operations there were never easy (see Chapter 4), the flat terrain and the ease with which enemy armour could be spotted meant that armour-killing became a particular specialism of the Flak regiments and battalions within Rommel's Afrika Korps.

Just how good the 88s were in this capacity became apparent to the Allies by mid-1942. Although at this stage of the war the USA had only been directly involved in the North African campaign for one month, the reports concerning the 8.8cm Flak as an AT weapon were sufficiently alarming for them to produce an intelligence report on the subject in June,

At a point in the Knightsbridge area [part of the Gazala Line defences], the 4th British armored brigade faced some 35 German tanks of the Mark III and IV type drawn up in line and obviously inviting attack. These tanks were supported by a battalion of anti-aircraft guns. The commander of the 4th Brigade refused to attack at all because of the presence of these guns on the battlefield. Slight firing occurred throughout the day. Towards evening the superior British tank force withdrew and the German tanks attacked after nightfall in a new direction. Their 88mm guns had checked the British all day and permitted Rommel to seize the initiative as soon as the British threat had vanished.[35]

RIGHT A Flak gun team deliver indirect fire on Allied positions during the fighting in North Africa in May 1943. *(Bundesarchiv, Bild 101I-787-0510-31/ Troschke/CC-BY-SA 3.0)*

The final observer report quoted in the document is emphatic in the status it gives to the 8.8cm Flak:

> *The greatest single tank destroyer is the German 88mm anti-aircraft gun. For example, on May 27th at 8:00a.m., Axis forces having enveloped Bir Hacheim [Bir Hakeim], a German tank force of sixty tanks attacked the British 22nd Brigade some distance to the northeast. The British moved to attack this force with 50 light and medium American tanks. It soon became apparent that this British force was inadequate and the Brigadier commanding ordered a second regiment of 50 tanks into action. In ten minutes the 88mm German AA guns destroyed 8 American medium tanks of this reinforcing regiment. All day thereafter, the British engaged the enemy half-heartedly and finally withdrew. Sixteen American medium tanks were lost in all. These sixteen fell victims without a single exception to the 88mm AA gun.*[36]

What is apparent from these accounts is that the battles in the desert in 1941 were teaching the Germans how to integrate the 8.8cm Flak weapons with armour and infantry. On the defence, the 88s would be positioned to create, in effect, anti-armour strongpoints that acted like breakwaters to split apart British armoured assaults. For example, when General Claude Auchinleck launched Operation Crusader – a further attempt to relieve besieged Tobruk – on 18 November 1941, his armoured units ran into three major 8.8cm strongpoints, at Bardia, Sollum and in the Halfaya Pass, each of around 12 guns with interlocking fields of fire. Halfaya Pass was already littered with victims of 8.8cm fire from previous attacks there, where the Germans used what would become a frequent tactic of withdrawing its tanks, as if in retreat, and thereby drawing the enemy armour on to

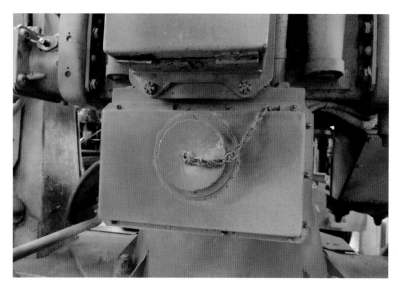

ABOVE The main electrical connection box on the 8.8cm Flak, set on the front of the pedestal. All electrical ports had similar chain-linked covers. *(Author/Axis Track Services)*

LEFT A Flak 18 lies wrecked in the North African desert, the debris around it indicating hits by artillery fire. *(AirSeaLand Photos/ Cody Images)*

ABOVE In this scene from the Mareth Line in Tunisia, April 1943, British troops inspect a spiked Flak 37 and its accompanying SdKfz 7 prime mover, abandoned during the retreat. *(AirSeaLand Photos/Cody Images)*

the 8.8cm weapons as the British gave pursuit. (In one attack by the 11th Hussars in June 1941, 11 out of 12 tanks in the first wave were destroyed by the 88s.)

Several factors made the 8.8cm Flak guns especially lethal to the Allied vehicles. First, because the gun crews were trained in hitting moving targets at long range (i.e. high altitude), they could engage the British, Commonwealth and later American tanks at great distance, typically as soon as they were visible. This meant that even if the Allied armour commanders did not fall into the tactical trap of being led on to the guns by a feigned German armour withdrawal, their tanks were still rarely safe. The only way to deal with the 88s at range was either to find dead ground that provided covered movement (often rare in the flat desert regions) or to put down counter-battery fire (mixed HE and smoke) that prevented the 88 gunners from either operating or from getting a clear view of the targets. The second issue was that the 8.8cm AP shell could smash through even the heaviest contemporary armour. For example, the most powerful of the British battle tanks in the Western Desert was the Churchill, which had 102mm (4in) armour on the hull front, 89mm (3.5in) armour on the turret front and hull side, and 76mm (2.9in) on the turret sides and rear. The textbook penetration figure for the Pzgr Patr 39/1 shell at 2,000m (2,188yd) was 127mm (5in), and at 1,000m (1,094yd) the figure was 159mm (6.3in). Although the penetration of tank armour is always dramatically affected by the angle at which the armour is presented, these blunt figures still reveal just how vulnerable even a heavy tank like the Churchill was to an 8.8cm strike.

The third, and arguably most important, factor in the rise of the 8.8cm Flak as an AT gun in North Africa, and indeed in the other theatres in which it fought, was its increasing integration into wider tactics. In April 1943, by which time the Germans were on their way to final defeat in North Africa, the Flak branch of the Wehrmacht

LEFT A Flak 18 emplaced in North Africa. The friendly vehicles in the distance give a good sense of the size of targets over this flat terrain. *(AirSeaLand Photos/Cody Images)*

released a manual called *Training Specifications for the Flak Artillery: Action and Waging War in Desert Combat*, in which they codified some important principles from the last two years of fighting. In the quoted passage that follows, note particularly the points made about tactical movement and about the synergy between the armour and the guns:

16. Heavy flak batteries that are to support the attack of their own tanks proceed with distances and intervals of 200 to 300 metres between the guns under the fire protection of light Flak artillery. The unit of fire is the single gun, commanded if possible by officers. The guns provide mutual fire support for each other. Single guns must always keep their fire on the enemy.

Firing is done from the chassis as a matter of principle, whereby the inexact firing situation is to be made up for by greater use of ammunition.

Frequent changes of position are necessary. Position changes of a minimum of 100 metres are usually fully sufficient to force the enemy to take new aim. Close cooperation between the accompanying Flak artillery and between one's own tanks is essential. The times of arrival and equalization of marching speeds and all movements must be agreed on precisely by the tank commander and the commander of the Flak artillery.

ABOVE A gunner carries a black-tipped armour-piercing shell during heavy fighting around Bir el Harmat in early June 1942. *(AirSeaLand Photos/ Cody Images)*

BREAKING OFF COMBAT AND WITHDRAWAL

17. Breaking contact with the enemy must take place earlier in the desert than in other theatres of war, since in coverless terrain it almost always takes place in view of the enemy.

18. Heavy Flak batteries are particularly suitable for providing cover for rearward movements on account of their range and firepower, since the enemy usually reacts strongly to disruptive fire when his artillery has marched forward.[37]

Here we see the 8.8cm Flak perceived in both offensive and defensive terms. Approaches to and withdrawals from the front line and the

LEFT Pulled by an SdKfz 8 heavy artillery tractor, a Flak 18 is drawn through the sandy terrain of North Africa in January 1942. *(AirSeaLand Photos/ Cody Images)*

ABOVE A Flak team at Bir el Harmat await Allied armour. Note how the gun is still on its wheels, ready for a quick redeployment after firing. *(AirSeaLand Photos/Cody Images)*

enemy positions had to be coordinated carefully to produce well-timed movements. By siting guns at 200–300m (656–984ft) intervals, the 88s were able to leave corridors of advance for the friendly tanks, while at the same time the independent command of each gun meant that every crew was able to support the German armour fully by identifying and engaging targets of opportunity as they arose. Note that these lines of 88s could be very extensive indeed; during the German Gazala offensive in May–June 1942, the Flak gun positions extended over 3km (1.8 miles), Rommel and his senior commanders having combed through rear areas and supply lines for any spare 88s available.

One of the most interesting comments here relates to advocating that the 88s fire from their wheels, for the sake of mobility and survivability. To be static for too long in the desert was to invite disaster, so the 8.8cm Flak crews stayed on their wheels to make them tactically responsive and also so that they could support the movement of armour over distance. For example, in the Gazala operation in early June, *Kampfgruppe Wolz* was formed principally out of AT and Flak battalions, and was led by Oberst Alwin Wolz, commander of the Flak Regiment 135 and later (on the Eastern Front) the 3rd Flak Division. In support of a push by 15th and 21st Panzer Divisions, the 88s of the battlegroup would work together in a mobile *Pakfront*, essentially performing a fire-and-manoeuvre action with half the battlegroup static and firing while the other half advanced forward with the tanks, then switching roles to allow the rearmost guns to move forward. (More about the *Pakfront* is detailed later in this chapter.) Subsequently a 4km (2.5-mile) long front of 88s covered the German Panzer withdrawal, and although these guns and their crews were hammered mercilessly by British artillery, which inflicted terrible casualties, the German gunners still managed to wipe out or immobilise no fewer than 50 British tanks.

Ultimately, neither the 88s nor the rest of the German forces were able to prevent the conquest of North Africa by the Allies, and the expulsion of the Wehrmacht from that theatre. The 8.8cm Flak guns had indeed inflicted a major physical and psychological scar on the enemy armour. In return, the Flak crews had been worn down by overstretched logistics (particularly affecting the flow of spare parts and ammunition) and by the mounting Allied fire superiority – one of the best defences against 88s was, it was realised, simply to pound the gun area to destruction with field artillery and air attacks.

LEFT The effects of a direct hit by an armour-piercing round on a Soviet T-34 tank, thousands of which were destroyed by 8.8cm Flak, KwK and Pak guns. *(AirSeaLand Photos/Cody Images)*

The Eastern Front

The ground war on the Eastern Front was fought on a scale, and with an horrific intensity, that is exceeded almost nowhere else in history. When Operation Barbarossa, the invasion of the Soviet Union, was launched on 22 June 1941, it brought the Wehrmacht into confrontation with an enemy of seemingly limitless scale in terms of men and vehicles. For the 8.8cm Flak crews, the conflict presented them with a number and array of targets that seemed scarcely believable. Furthermore, although the Soviets had some excellent individual tank types – especially the legendary T-34 – in the first year of the war these assets were almost invariably handled with tactical crudity and a seemingly oblivious attitude to minimising losses. It is not uncommon to see photographs of 8.8cm guns with 40+ tank kills chalked up on their shields or barrels. Between 22 June and 15 September, I Flakkorps alone wiped out 3,000 armoured vehicles.

The sheer volume of vehicular targets, which at times came surging on in vast uncoordinated waves, meant that new tactics had to be developed to cope with them. At first, the 88s operated largely as they had been doing in North Africa, with individual guns picking out and destroying individual targets at will. This approach, however, struggled to cope with truly large numbers of assaulting tanks; numerical strength in one part of the front might overwhelm the German defences there, leading to a Soviet breakthrough and the 88 crews' nightmare – being outflanked. One incredible description of just how close-quarters the tank vs 8.8cm Flak battle could get comes from one Feldwebel Karl Hass, who encountered what

BELOW The shield of this 8.8cm Flak 18 on the Eastern Front proudly displays a wide variety of kills, from enemy aircraft to bunkers. *(AirSeaLand Photos/Cody Images)*

ABOVE During the heavy fighting near Voronezh, Russia, in 1942, a Flak 18 gunner uses a *Entfernungsmesser 34* (EM34) 0.7m coincidence rangefinder to spot for targets. *(AirSeaLand Photos/Cody Images)*

BELOW Another scene from the action around Voronezh. With vehicles burning on the horizon, the gun crew spot for their next targets. *(AirSeaLand Photos/ Cody Images)*

from his account appear to be T-34s and the heavy KV-2 during one battle:

More and more tanks came at us. One advanced to within 60 metres of the gun positions and we fired into them. Several gunners were wounded. Another direct hit put a whole gun crew out of action.

We targeted a second tank. Then a shell crashed against the left shield of our 88. We were all rather stunned, as if after a bad fall. The gun leader, who had not been hit, was the first to see a gun barrel firing from behind a heap of straw: a camouflaged tank! It had come to within 100 metres of us. Then finally the shot was fired and glanced off the monster. The second shell was loaded. Meanwhile the tank had come to 30 metres. The shot flew out of the gun and – in the turmoil, it went past! 'It's all over now,' we all thought. 'Now he'll drive up and roll right over us.' The steel Colossus rolled over the truck and slammed into a 20mm and our heavy Flak gun. It struck so hard that the 88 was slammed backwards into the Flak gun standing behind it. And the 52-ton tank was here. Meanwhile the 34-ton tank had driven up to three other trucks, tried to roll over them, climbed up steeply over the first one and began to tip. The truck tipped over and was crushed. Meanwhile the 52-ton tank had rolled over two 105mm guns and a row of trucks behind our guns, then turned back and, firing steadily, made its way back.[38]

A general point to be made about this episode was that during the fighting on the Eastern Front it became progressively common for 8.8cm Flak crews to fight at close quarters with the enemy, both armour and infantry, such was their prevalence at the harshest points on the front line. The gun crews began to invest more seriously in plentiful supplies of small-arms weapons and ammunition, plus hand-held AT weaponry, hanging their survival on more than just the big gun. Owing to the regularity with which Flak crews fought with heroism against enemy ground troops and armour, in March 1942 a new decoration was introduced: the Ground Combat Medal of the Luftwaffe. Also, the fact that official Luftwaffe guidelines came to give 600m (656yd) as the maximum range at which to engage T-34s, indicates the close proximity with the enemy at which the battle was fought.

Nor was it just armoured vehicles the 88s were employed against. There was a notable expansion of roles for the Flak guns on the Eastern Front – working as direct-fire fortification busters against Soviet field positions and against major emplaced works, such as the vast fortress complex at Sevastopol; engaging river boats attempting to sail along some of the mighty eastern waterways, or against naval vessels operating along the Baltic coastline; delivering HE barrages on Soviet troop concentrations; as well as performing the AA duties for which they were designed.

With such pressures on the Flak crews, fire had to be better coordinated and applied, hence the aforementioned *Pakfront* tactic was implemented from late 1941. In this arrangement, up to ten 88s were arranged under a single commander, who essentially took responsibility for fire direction. The commander would assign targets to each gun, and timings at which they had to fire. Although this approach sounds more cumbersome, it actually meant that the Flak units could respond to threats more effectively, and could also launch surprise ambushes in a more coordinated and lethal fashion. In fact, such was the success of the *Pakfront* tactic that the Soviets themselves copied it.

Yet throughout the innovation and the almost endless armour kills, there were several problems mounting for the 8.8cm Flak crews, especially as the expected German victory became problematic (late 1941), then uncertain (1942–43), then failed (1944–45). Across the Eastern Front, the 8.8cm Flak guns were increasingly acting in a 'fire brigade' capacity, being channelled into support for Army troops at locations of greatest threat and combat intensity. This meant that 88s, with soul-destroying frequency, found themselves bearing the brunt of anti-armour combat in sectors that were collapsing, the gun crews often fighting either to hold tactically worthless positions deemed crucial by Hitler, or providing increasingly lonely cover fire to retreating forces. In either case, numerous guns ended up destroyed and the crews killed. By 1943, German war manufacturing was growing, but the pressures for production of every weapon system meant that there were rarely enough 88s at the front line. The Soviets, meanwhile, were putting more and more armour on to the battlefield – total Soviet armoured vehicle production for the war years was an astonishing estimated 120,000 – and they were also using that armour with improved tactical intelligence. Thus the Flak crews could only knock out so many tanks before sheer weight of numbers and fire brought disaster. This situation was worsened by the fact that with casualties running at high levels among the gun teams, the 8.8cm Flak weapons were increasingly manned by inexperienced crews combed from the Reich or the rear areas,

ABOVE Operation on the Eastern Front. Note the heavy build-up of ice and snow on the barrel; this had to be removed regularly or it could affect the weighting of the gun. *(AirSeaLand Photos/Cody Images)*

BELOW The collar around the Flak 18 barrel, for connecting with the muzzle rest. *(Author/Muckleburgh Collection)*

LEFT This Flak gun is seen on the outskirts of Stalingrad in September 1942. In this famous battle, the 8.8cm Flak guns were heavily used for delivering preparatory bombardments of the city. *(AirSeaLand Photos/Cody Images)*

ABOVE A German soldier inspects the puncture hole of an 8.8cm armour-piercing round in the side of a T-34 tank. Note the metal spall radiating out from the hole. *(AirSeaLand Photos/Cody Images)*

LEFT Photographed during the first month of Operation Barbarossa in the Soviet Union, these men of the 79th Mountain Artillery Regiment take a break from firing on Red Army bunkers near Galusinzy. *(AirSeaLand Photos/Cody Images)*

soldiers who either learned on the job *in extremis* or were quickly killed or wounded.

A further complication of the Flak crews in AT roles was that of command. Flak guns and teams were deployed to wherever it was felt they were most needed, but in many cases they sat in an uncertain command relationship between their Luftwaffe parent unit and the Heer unit they were supporting. It sometimes became unclear whose line of command they were representing (especially if the gun teams were being moved rapidly between sectors), and to make matters worse the Army commanders

LEFT A battery of 88s participates in the German bombardment of Sevastopol, Crimea, in 1941–42. *(AirSeaLand Photos/Cody Images)*

LEFT Eastern Front, 4 November 1942. A Flak 18 crew set up their weapon behind wrecked buildings. Note the cover over the breech to protect the vital mechanical parts from the ingress of dust and dirt. *(AirSeaLand Photos/ Cody Images)*

might sometimes forget to include the Flak units in their wider commands. Clarifications were issued in doctrinal documents, such as this one in September 1942:

3. The Flak units in action in an area of operations are instructed to cooperate with the command posts of the army, as a rule, as follows:

Flak corps with army groups
Flak divisions with army upper command
Flak regiments with army corps
Flak units with divisions.

4. The command posts listed above must, by making command conditions fully clear, avoid turning cooperation into subordination, or loss of cooperation through unduly loose communication as to the mutual target.

5. The regulation of command conditions must extend not only to one's own Flak forces, but should also extend to the Flak artillery of the army and the Waffen-SS.

Only the tactical subordination of these units under the command posts of the Flak artillery guarantees unified action, conserves forces and guarantees focal-point formation and success.[39]

Between the lines, we can almost sense the shrill complaints and arguments about front-line confusion over orders and command, more than occasionally terminating in dire physical consequences for gun crews. It is hard to tell whether such guidelines and clarifications actually had much effect in the snowballing collapse of German fortunes on the Eastern Front between early 1942 and the defeat at Stalingrad, and the final battles for the Reich in 1945. In 1944, furthermore, the Army was given greater authority, direct from Hitler himself, to appropriate the Flak guns to whatever purpose was required.

By 1945, and the final battles for the Reich, the Flak batteries were often in a parlous state.

BELOW **A Flak gunner observes through his ZF 20E optical sight and makes adjustments using the traverse mechanism.** *(AirSeaLand Photos/ Cody Images)*

137

AT WAR – ANTI-TANK OPERATIONS

ABOVE Amid the fighting near Stalingrad in September 1942, an 8.8cm Flak gun moves forward to support operations by German armour. *(AirSeaLand Photos/Cody Images)*

Logistics were beginning to crumble in the Wehrmacht, resulting in serious shortages of ammunition and spare parts. The transfer of city-based AA units to field-based work (see Chapter 5) resulted in many of the /2 guns being crudely mounted on trucks, horse-drawn wagons and anything else that came to hand. Nevertheless, the 88s still inflicted a heavy toll on Soviet armour, right up until the very last acts of the war.

Fall in the West

By the time that the war in North Africa moved into Sicily and then mainland Italy in 1943, the British, American and other Western Allied forces were already painfully aware of the capabilities of the 8.8cm Flak guns, and thus proceeded with caution when it was known that 88s were in the vicinity. For the German gun crews engaged in anti-armour work, the

RIGHT The ZF.20 optical sight unit. The dial seen top right is the azimuth deflection knob. *(Author/Axis Track Services)*

FAR RIGHT The elevation micrometer on the ZF.20 sight. *(Author/Axis Track Services)*

western European theatre presented them with a specific set of threats and opportunities. One key problem was that of range limitations presented by terrain. A US intelligence analysis of 8.8cm range capabilities from November 1943 explained that:

> The 88-mm guns can open fire on armored vehicles at 2,500 yards with fair prospect of success, but are most effective at ranges of about 1,000 to 1,500 yards. They may fire at ranges of as much as 4,000 yards, if other and more inviting targets are not available. With the aid of a forward observation post, 88s sometimes engage such targets as troop concentrations at ranges of as much as 6,000 yards.[40]

The assessment is correct in noting that the short ranges were optimal for the Flak guns as

ABOVE An emplaced Flak gun in the Italian theatre, the shield heavily covered by foliage. The rings on the barrel suggest a large number of combat kills. *(Bundesarchiv, Bild 101I-304-626-05A/ Lüthge/CC-BY-SA 3.0)*

LEFT In a strange ceremony, two members of a Flak crew in North Africa are married remotely to their sweethearts back in Germany, while the other gunners take photographs. *(AirSeaLand Photos/ Cody Images)*

LEFT This destroyed Pak 43 was dug in along a rural French road, using the hotel in the background – itself the victim of shell strikes – as partial cover. *(AirSeaLand Photos/Cody Images)*

CENTRE The turret of a Tiger II tank, here at the wrecked Henschel plant in April 1945, shows the extreme length of the KwK 43 gun – 6.24m (20ft 6in). *(AirSeaLand Photos/Cody Images)*

anti-armour weapons. In convoluted western European terrain, however, even getting an unobstructed view of 1,000m (1,093yd) could be problematic, depending on the country and the landscape. The Netherlands, for example, offered decent flat expanses over which the Flak guns could fire at extended range, depending on features such as trees and buildings. In rural Italy or Normandy, by contrast, combat ranges could be forced by terrain down to just a few hundred metres. Indeed, in photographs from the European Theatre of Operations (ETO) it is not uncommon to see Flak 8.8cm guns set up orientated down a short stretch of village street or urban road, or seated camouflaged in the hedgerows bordering a field, waiting to ambush the enemy at close quarters rather than strike out at him over range. To deliver such ambushes effectively, it was imperative that the 88 guns had the supporting presence of infantry and faster-firing guns, such as 2cm Flak, another AA weapon that was a highly effective tool against ground targets.

Intelligent positioning could, of course, give the 88 gunners a longer perspective over the terrain. In Italy and in hilly areas of northern Europe, guns might be located on high ground, as long as it was suitable for quick egress when necessary. A bit of elevation, especially from a vantage point that overlooked a likely avenue of Allied advance, could turn a valley or an expanse of open fields into a killing ground for the 88s. Whatever the range, camouflage was also essential, especially so in areas where the Allies had air superiority

LEFT The crew of this Flak 18 pose next to the weapon, the front end of the barrel blown off by what was obviously a catastrophic malfunction. *(AirSeaLand Photos/Cody Images)*

RIGHT Oerlikon cannon-armed British troops keep watch from a captured German coastal emplacement in Normandy, a now useless Pak 43 still menacing from the embrasure. *(AirSeaLand Photos/Cody Images)*

(basically the whole of the Western Front from late 1944). The gunners would typically use local vegetation and camouflage netting, building it up around the gun to break up the weapon's visible lines and angles.

One location in which the visible distance was rarely a problem was in coastal positions. The 8.8cm Flak guns were heavily deployed

BELOW The Flak gun recoils heavily as it engages ground targets on the Eastern Front. Note how the gunners form a queue to the gun with fresh shells. *(AirSeaLand Photos/Cody Images)*

ABOVE A US Sherman tank passes an abandoned 8.8cm Flak gun. The Sherman crew have up-armoured themselves by packing sandbags on the front hull. *(AirSeaLand Photos/Cody Images)*

around Europe's coastlines, not just for the anti-aircraft defence of ports and other coastal locations (see Chapter 5), but also for providing firepower against enemy shipping and against potential amphibious landing forces. In March 1943, the Western Front coastal Flak batteries received special orders regarding their roles, positioning and tactics should there be an attempted Allied landing. It was explained that the guns needed to be positioned to deliver both direct and flanking fire against the invading enemy, being able to shoot across the breaches and engage targets in the sea at both ebb and flow tides. They also had to exercise target-fire discipline, avoiding slugging it out with heavy offshore naval vessels and instead concentrating on those craft that provided the most immediate threat – amphibious landing craft and assault boats, plus, in the air, any aircraft deploying parachute forces. One paragraph emphasised that if the enemy did manage to get ashore, the Flak crews

ANTI-TANK TACTICS: TEXT EXTRACTS FROM A CAPTURED GERMAN DOCUMENT (OCTOBER 1944, WESTERN FRONT)

Don't split up antitank units, give them definite tasks in combat, maintain close liaison with the infantry, set up antitank nests under unified command, employ self-propelled companies in mobile operations – these are some of the antitank tactics outlined in a recently obtained German document. The translation from the German follows:

The tendency to split up antitank units completely, to have a proportion of antitank firepower everywhere, is wrong. The smallest unit permissible is the half-platoon (two guns), except for defense of streets for which less may be employed.

Companies in their entirety, or at least whole platoons, should cover likely tank approaches. To use a single antitank gun is to invite destruction. Other terrain over which tanks might approach will be covered by mines, obstacles, or tank-destruction detachments.

Antitank units will normally be in support; they must be given definite tasks and allowed to make their own tactical dispositions.

Engagement of even worthwhile infantry targets must be the exception rather than the rule. Such employment is limited by lack of mobility, by the bulkiness of the gun as a target, by the sensitivity of the barrel which is subjected to great strain, and finally by the small issue of high-explosive shells. In addition, accuracy diminishes with bore wear.

On the move, regimental antitank companies are normally distributed throughout march groups by platoons – one platoon always with the advance party. No heavy antitank guns [should be] with the point, as too much time is needed to bring them into action. Divisional antitank battalions are normally brought forward as a body. In assembling, locate in areas from which the final movement can also be protected; local protection [should be] by machine guns. Positions for antitank guns not immediately employed will be reconnoitered and prepared. Antitank warning arrangements must be made by the officer commanding the antitank unit detailed for local protection. Advantage will be taken of unexpected gains of ground to push forward the antitank defenses.[41]

were expected to fight in 'close combat', until reinforcements arrived.

The coastal 88 crews on the Western Front did indeed get to see action on several occasions. In Italy, 8.8cm Flak guns warranted a personal mention from Generalfeldmarschall Kesselring for their invaluable role in keeping Allied forces choked up on the Anzio bridgehead, following the landings there on 22 June 1944. Every time the British and American tanks tried to push inshore, they were cut up by 88 fire, such that a breakout wasn't achieved until the following May.

The 88s also added their weight of fire to the German attempt to repel the Allied D-Day landings at Normandy on 6 June 1944. The WN 83 strongpoint at Les Perruques, for example, held 12 8.8cm Flak guns, which not only engaged overflying C-47s that were dropping paratroopers over Normandy, but they also delivered heavy fire over on to Omaha Beach, contributing to the terrible American casualties experienced on that beach in the first day of Operation Overlord. The battery was eventually suppressed on 9 June by troops from the US 2nd and 5th Rangers, the 116th Regiment of the 29th Division and the 83rd Chemical Weapons Battalion (Heavy Mortars), with the support of US field artillery.

Following the D-Day landings, the fighting developed into a tortuous close-quarters wrestle in Normandy's famous *bocage* countryside. Although this was not really the environment best suited to the 88, there is no doubt that – whether as Flak guns, Pak weapons or mounted on tank-destroyers or tanks – they still imposed frequent and heavy losses on the Allies. A selection of Canadian after-action reports from the post-D-Day fighting illustrates the point:

The first three self-propelled artillery vehicles of 14th Canadian Field Regiment had just pulled off into a field at the right of the road

BELOW US troops inspect an abandoned Flak 36 or 37, emplaced in dense undergrowth on the Western Front in 1944. *(AirSeaLand Photos/ Cody Images)*

and the detachments were preparing the gun position, when a hidden 88mm gun fired on them hitting all three vehicles in a matter of less than a minute. They burst into flames and the huge quantity of ammunition carried on them (in addition to 105mm ammunition, this included extra ammunition and mines for other arms) commenced to explode. The detonations lasted over an hour and made the area an exceedingly dangerous one. As a result of this incident the tanks were loath to leave cover and advance up the grassy field leading the infantry.[42]

The 88mm still appears to be the enemy's most formidable anti-tank weapon. It is most effective against Shermans at ranges up to 3,000 yards. All his anti-tank guns are very well sited and full use is made of dummy guns as decoys. Snipers have been found dug in well forward of anti-tank gun positions. These force crew commanders to close down and limit their field of vision.

Tanks encountered so far have been Marks III and IV and Panthers. Tigers have also been reported but not confirmed. Mark IVs have been captured dressed up with light spaced armour to resemble the Tiger. The deception is hard to detect. It can be stated definitely that we are outgunned both by the long 75mm and the 88mm.[43]

The second of these accounts provides an insight into the cooperation between Flak 8.8cm guns and the surrounding infantry, the latter providing suppression via their small-arms fire to obstruct Allied situational awareness. In many cases, the Allied tank crews would know nothing about the presence of a well-concealed gun until tanks began to explode.

Much of the fighting for Normandy was at the small-unit level, with small numbers of Flak guns deployed to meet an immediate tactical threat. Yet the Germans also used the 88s in more substantial defensive deployments during the battle for France. Within the fighting for Caen, for example, the Heer and Luftwaffe pooled resources to field hundreds of 88s positioned along the Bourgébous Ridge. These guns, and other AT assets, were responsible for destroying some 400 Allied tanks following a concerted armoured push against Caen on 18 July 1944. One Flak gun alone took out 12 British tanks. As with the Soviet forces out east, it was only the great industrial capacity of the Allies that enabled them to ride out such losses.

Having acknowledged, once again, the undoubted killing capacity of the 88, we could conversely argue that on the Western Front, in many ways, the 88s ultimately found themselves even more on the back foot than on the Eastern Front. In the days following the D-Day landings, the Allies established virtual air superiority over the whole theatre, with fighter-bombers making relentless predatory ground-attack sorties against any field-grey target below. Large numbers of Flak guns and

BELOW The muzzle of a Pak 43 – apparently damaged by enemy fire – near the Utah Beach landing zone in Normandy. *(AirSeaLand Photos/Cody Images)*

BELOW RIGHT The hinged base section of the 8.8cm Flak outriggers. *(Author/Axis Track Services)*

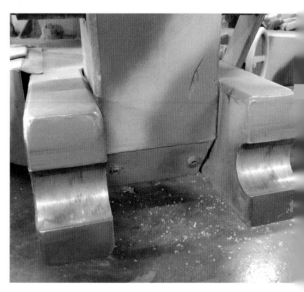

ABOVE Generating a cloud of dust (another good reason for fitting the muzzle cover in transit), a Flak 18 is drawn into a firing position by an SdKfz 7.

crews were wiped out on the roads to the front, especially as the Germans attempted to rush large numbers of AA 88s forward from their positions around Paris. Allied forward air observers would also work closely with infantry and armour units, using radio communications to give precise coordinates of any 8.8cm gun positions once they revealed themselves. Air attacks could come within minutes of the radio transmission being released.

The close confines of much of the northern European landscape meant that the 88s were also exposed to aggressive ambush tactics from the Allies, as this after-action report from the US 37th Tank Battalion, 4th Armored Division, illustrates:

The Battalion moved from Bain de Bretagne at 1715 and pulled into bivouac four (4) kilometers Northeast of Janze (158398 – vicinity Amanalis) at 1925. Just before going into Janze, the Battalion shot up several trucks and destroyed several machine gun nests. AT Amanalis, 'D' company and 'B' Company were sent North to cover road junctions and to prevent enemy troops from entering or leaving Rennes, which was to be assaulted by a CT of 8th Infantry Division. 'D' Company covering the RJ's Conrad Mueller of 'D' company, who commanded the outpost at RJ at 164440, was informed that a column of about 500 infantry and 2 towed 88s were coming down the road. Lt. Mueller

RIGHT Local people mill around a Flak 18 in its position in Tunisia in February 1943. *(AirSeaLand Photos/Cody Images)*

BELOW The barrel of a KwK 43 stretches from a Tiger II out over a French road, the tanks hiding from aircraft. The two British soldiers are POWs. *(AirSeaLand Photos/Cody Images)*

laid the 105mm assault gun attached to him and as the first gun rounded the corner blasted it and its crew. The Jerries – never known to be particularly bright – hesitated for several minutes and then dismounted men and cleared away the wreckage of this gun.

The Germans then tried to push their second towed gun around the corner, only to have it meet the end of its predecessor.[44]

The comment about the German Flak crews' level of intelligence can largely be put

down to nationalistic bravado. It must also be acknowledged, however, that many of the 88 gun crews in France were relatively inexperienced, especially in anti-armour warfare, compared to their comrades on the Eastern Front. Rather than lacking acuity in this encounter, the Germans might simply have been lacking tactical awareness.

One final point about the fate of the 88s on the Western Front was that their enemies here were, in general, less willing to sustain the levels of casualties suffered by the Soviets. This factor, plus a generally high level of officer training, made them an increasingly wily opponent once they had gained relevant combat experience. At the same time, they brought with them truly enormous levels of firepower, especially in terms of armour, artillery and aircraft. A tactical sophistication backed continually by the means to assert fire superiority made the Western Front a costly theatre for the German 8.8cm guns and their crews. One only need look at the photographs of the destruction unleashed on the Germans in the Falaise Gap, around Monte Cassino in Italy or around cities such as Cologne, to understand that the life of an 8.8cm gunner was a short and brutal one between 1944 and 1945.

One last comment needs to be made about the 8.8cm Flaks guns as AT weapons. That they were good at tank killing, there is no doubt. But neither should we give the 88 a position superior to many other AT guns, including those of the Allies. In many ways, the Americans, British and Soviets all fielded AT weapons that were the equal of the Flak 18/36/37/41. The British 17pdr and 3.7in Mk 3 and the American 90mm M1 could match and even outstrip the performance of the 8.8cm weapons, in some cases firing heavier shells to longer ranges, and with armour penetration capabilities exceeding the Flak guns at every range. The Flak weapons, however, were available to the German forces right from the beginning of the Second World War, and in significant numbers. Develop any good weapon, mass produce it and place it in well-trained hands, and it will have a powerful influence on the battlefield.

BELOW Operating the 8.8cm Flak gun in anti-tank mode was exhausting for the loader, as he had to hoist each shell up high to reach the horizontally aligned breech. *(AirSeaLand Photos/Cody Images)*

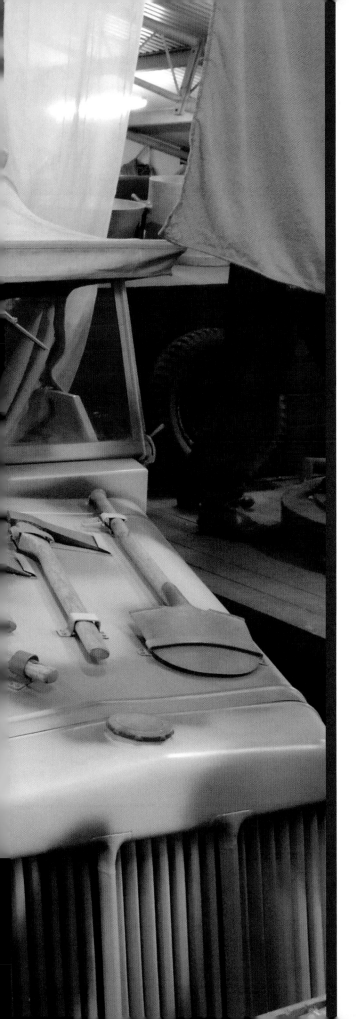

Chapter Seven

Transportation and maintenance

The 8.8cm Flak gun depended on its half-track prime mover to pull it around the battlefields of the Second World War. It also relied on a crew who were prepared to care for the gun mechanically on a daily basis, performing all the necessary maintenance to ensure that it did not fail them in combat.

OPPOSITE A front view of a well-restored SdKfz 7. The tools attached to the bonnet would have been essential for creating an effective gun position. *(Author/Axis Track Services)*

The carriage system of the 8.8cm Flak gun has already been explained in previous chapters. It would be remiss, however, not to spend some time describing the primary means by which the gun moved around the battlefront. Here we will briefly explain the principal 88 prime mover, the SdKfz 7 half-track, noting other vehicles that drew the gun around the battlefront. This mechanically minded chapter then goes on to look at the practical care and maintenance of the gun itself, activities that filled most of the working life of an 8.8cm gun crew.

Half-track prime movers

The SdKfz 7 was not the only means by which the 88s travelled around the battlefronts of the Third Reich. As we have seen, some were mounted on railway wagons, providing a certain degree of mobile AA defence around German cities. We must also acknowledge the contribution of improvisation to manoeuvring the guns – doubtless the 88 was shackled to all manner of vehicles when emergency or necessity warranted. Yet it was the SdKfz 7 that was the principal transportation for the 8.8cm Flak.

The design process for the SdKfz 7 began in the late 1920s, coming out of an Army requirement for an artillery tractor and general prime mover. It was developed by Krauss-Maffei AG and limited production began in 1933, with full-scale output getting under way in 1939 through several companies, including Borgward, Daimler-Benz and Saurer Werke. In total, 12,187 of these vehicles were manufactured during the war years.

The major layout decision behind the SdKfz 7 was to opt for a half-track design. This was a sensible move, for the tracked portion of the vehicle, with seven overlapping road wheels each side, gave it the traction and power to pull a heavy artillery piece over the most compromised of roads and fields, while the front wheels provided the fine steering that was much appreciated when trying to position a gun with some precision. The SdKfz 7 was

BELOW Two SdKfz 7s, both fitted with the protective canvas awnings, and their 8.8cm guns parked under the cover of trees. *(AirSeaLand Photos/Cody Images)*

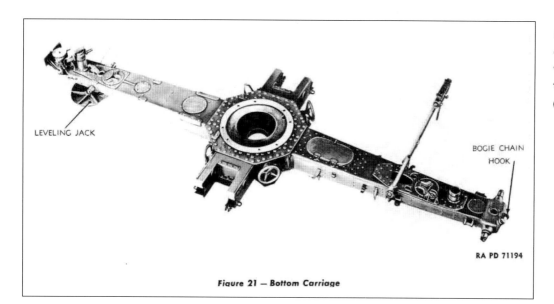

LEFT The fixed arms of the bottom carriage, which connected to the two bogies. *(US War Department)*

Figure 21 — Bottom Carriage

a substantial vehicle, measuring 6.85m (22ft 6in) in length, 2.35m (7ft 9in) in width and a height of 2.62m (8ft 7in). With an overall vehicle weight of 11.53 tonnes (12.7 short tons), it could pull a load of 8,000kg (17,637lb), the power delivered by a Maybach HL 62 TUK six-cylinder petrol engine via a six-gear (five forward, one reverse) transmission. (The more powerful HL 64 was fitted to production models from 1943.) Given its purpose, it was not the most sprightly of vehicles, but it could still maintain a constant road speed of 50km/h (31mph), and it could travel 250km (160 miles) on a full fuel tank.

LEFT The rear of a restored SdKfz 7, showing the tailgates and hinged sides, the bench seating for gunners, plus racks for small arms. *(Author/ Axis Track Services)*

SDKFZ 7 SPECIFICATIONS

Weight	11.53 tonnes
Length	6.85m (22ft 6in)
Width	2.35m (7ft 9in)
Height	2.62m (8ft 7in)
Crew	One driver
Passengers	11
Engine	Maybach HL 62 TUK, six-cylinder petrol (99kW/133hp)
Transmission	Six gears (five forward, one reverse)
Power/weight	9kW (12.1hp)/tonne
Payload capacity	1,800kg
Suspension	Torsion bar
Fuel capacity	215 litres
Operational range	250km (160 miles) on roads 120km (75 miles) off-road
Speed	50km/h (31mph) on roads

RIGHT The driver's compartment for the SdKfz 7. *(Author/Axis Track Services)*

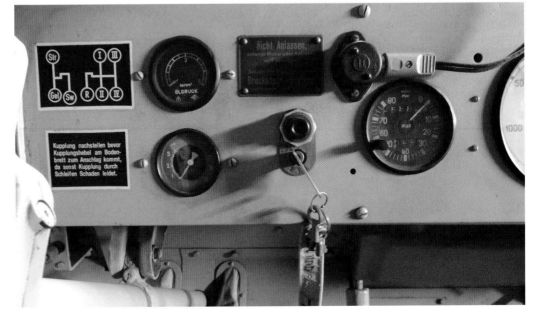

RIGHT A close-up of some of the driver's controls in the SdKfz 7. Note the diagram top left, showing the gear positions. *(Author/Axis Track Services)*

The virtue of the SdKfz 7 for the Flak 8.8cm was its rear capacity. There was storage space for ammunition, tools and other equipment behind the front cab, and bench seating for an entire gun crew in the back cabin (the driver and gun leader sat in the front). Weather protection came from a canvas tilt stretched over a wire frame.

The SdKfz 7 not only towed 8.8cm guns, but also a variety of other ordnance, particularly 10.5cm and 15cm pieces. Nor, as noted earlier, were the 88s towed by the SdKfz 7 alone. A study of the photographic sources show a number of other vehicles regularly making an appearance as 8.8cm transporters. The SdKfz 8 is one example. This Daimler-Benz vehicle, despite the apparent logic of the numbering, was actually developed before the SdKfz 7, again as an artillery prime mover. It had the half-track layout, but was a more powerful vehicle than the SdKfz 7, able to tow guns up to 12 tonnes (13.2 short tons) in weight, or 14 tonnes (15.4 short tons) in its most powerful DB 10 variant, which had a Maybach HL 85 TUKRM 8.5L 12-cylinder water-cooled petrol engine, delivering 138kW (185hp). For this reason, the SdKfz 8 is often seen pulling the heavier German artillery pieces, such as the 21cm Mörser 18, the 15cm Kanone 18 and the 10.5cm Flak 38, but it regularly drew the standard 8.8cm Flak 18/36/37 guns and the heavier 8.8cm Flak 41. A total of around 4,000

LEFT A fully armoured SdKfz 7 *Bunkerknacker* ('Bunker Cracker'), a relatively rare conversion of the standard vehicle, pulling a shielded Flak 18. *(AirSeaLand Photos/Cody Images)*

BELOW Another view of the SdKfz 7 *Bunkerknacker* vehicle. The crew were wrapped in a 14mm (0.55in) thick armoured cupola. *(AirSeaLand Photos/Cody Images)*

were produced, as opposed to the 12,187 of the SdKfz 7s.

Other types of towing vehicle do appear in the photographic records, such as the Büssing-NAG BN 9 and the Hanomag SS-100 wheeled tractor, but these instances are relatively rare and likely reflect improvisation rather than recommended practice.

Inspection, troubleshooting and maintenance

Like any complex weapon system, the Flak 8.8cm required meticulous daily maintenance to support its reliability and functionality. This maintenance depended utterly on the self-discipline of the crew and the command discipline of the gun leader, for it was all very well to clean and lubricate a gun daily at a peacetime barracks, but it was another matter entirely to do so when exhausted, cold and mentally shattered in a combat zone.

On deployment, the Flak team would have a set programme and schedule to follow that would cover all the necessary maintenance functions for both the gun and also the towing vehicle. A service schedule issued to an 8.8cm

ABOVE A view of the half-track track links of the SdKfz 7, and part of the torsion bar suspension. *(Author/Axis Track Services)*

BELOW The towing connection at the rear of the SdKfz 7. The vehicle had a towing capacity of 8 tonnes (8.8 short tonnes). *(Author/Axis Track Services)*

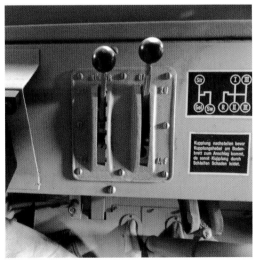

BELOW Set to the right of the driving position, the opening and closing levers for the SdKfz 7's radiator cooling slots. *(Author/Axis Track Services)*

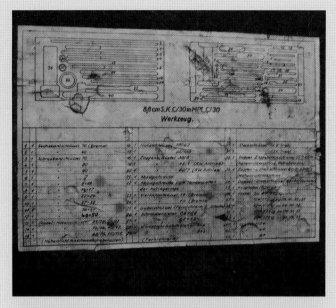
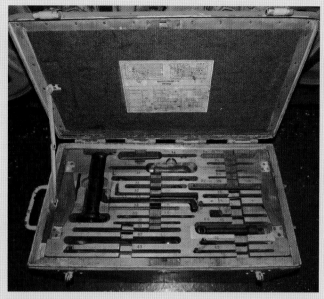

ABOVE A box of tools for essential maintenance and repair on the 8.8cm Flak gun, the guide to each individual tool pasted inside the lid of the box. *(Author/Axis Track Services)*

Flak unit in September 1941 on the Eastern Front had the troops rise at 05.30hrs, go through wash-up and breakfast from 06.00hrs, then fall in for roll-call and work at 06.30hrs. The maintenance work would then split into two main groups. One group would attend to the tow vehicle, specifically checking:

- Fuel levels
- Water levels
- Oil levels
- Motor function
- Valves
- Carburettor and air filter
- Tyre pressures and chains (if fitted)
- Vehicle lubrication
- The tightness of any pins, screws and bolts.

The ammunition loaders would also clean out the vehicle, both the driver's cab and the rear compartment, plus any vehicle logs would be updated. (One continually impressive discovery of military researchers is how diligently German vehicle crews tended to keep their logbooks, even under the most extreme combat pressures in the very last days of the war.)

For the gun crew, their inspection priorities were as follows, with some extra notes provided here of key maintenance and repair activities:

ABOVE On the Eastern Front in deep winter, a heavily clad Flak team fire at Red Army targets, the barrel almost perfectly level with the terrain. *(AirSeaLand Photos/Cody Images)*

Bore

The bore had to be inspected for any signs of rust, significant wear or damage, and for any build-up of dirt, debris and fouling (the latter usually from unburnt powder). At best, such issues could interfere with the spin stabilisation of the shell, making the gun less accurate; at worst, a defective barrel or bore liner could lead to the ultimate disaster, the explosive detonation of a shell before it had left the muzzle. After each firing, the gun required thorough cleaning. This was typically performed by washing out the barrel with a mixture of water and sodium carbonate, then drying it out with a bore sponge and rags, before lightly oiling the bore with engine oil. If the bore showed signs of damage or wear that threatened the integrity of the weapon, then the bore liner might have to be removed and a section replaced. Minor rust would be treated in situ, but more serious developments would also likely necessitate the replacement of bore liner components.

Removing the bore liner involved, in very broad outline, first detaching the barrel from the cradle and setting it firmly on two blocks. Then a drive plug was inserted into the bore at the muzzle end of the barrel, and a ram used to push against it, thereby pushing out the inner bore tube from the rear of the gun.

Breech mechanism

The breech mechanism was the mechanical heart of the gun, and thus like the barrel it needed diligent daily attention to ensure it functioned properly under its extreme workload. The mechanism would be cleaned after every period of firing, with propellant residues removed with solvents and all moving parts lightly oiled. (Note that on the Flak 8.8cm guns the lubrication points and oilholes would generally be marked with red paint to make them easier to identify.) After cleaning, the gunner would then operate the breech-block repeatedly, opening and closing it to detect

LEFT Breech components, such as the cocking lever seen here, were very exposed to the elements, and needed regular and proper lubrication to prevent rust impairing their function. *(Author/Axis Track Services)*

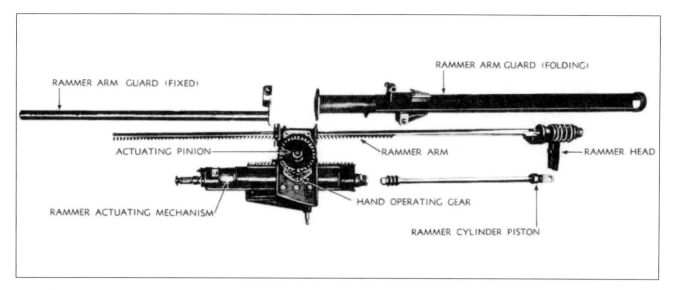

ABOVE An exploded view of the automatic rammer assembly, especially useful when the gun was firing at high elevations. (US War Department)

any stiffness, catching or other interference that would indicate a problem with either the breech or the breech-block, particularly looking out for any damage such as dents, burrs or pitting. If the breech mechanism was not functioning properly, it would be disassembled, parts replaced as necessary, lubricated, then reassembled.

Firing mechanism

The firing mechanism had to function crisply, with a fast lock time (the time interval between pressing the firing lever through to the shell primer detonating) and a reliable impact of the firing pin on the base of the cartridge case. To test its operation without firing the weapon, the gunner would pull on the firing lever, then open the breech and note whether the firing-pin mechanism had been cocked during the breech opening. He would then close the breech, release the firing pin, then cock the firing-pin mechanism using just the cocking lever assembly. If the firing mechanism failed either to cock or fire during this dry run, then the mechanism would need to be disassembled and parts replaced as necessary.

Traverse and elevation

On a daily basis, the elevation and traverse components were tested for smooth operation, the operator noting if there were any sticking points or unwanted play in the mechanisms.

LEFT A US soldier demonstrates the procedure for cocking the automatic rammer, rotating the rammer crank in an anti-clockwise direction. (US War Department)

157

TRANSPORTATION AND MAINTENANCE

Most detected faults could usually be remedied by thorough cleaning and proper lubrication, but any major mechanical tasks, such as the repair of drive teeth, would have to be done by senior engineers. Note that if the elevation wheel proved difficult to operate, the problem could also be that the equilibrators were not compensating for the weight of the gun properly. Repairing this problem might involve disassembling the equilibrators and re-tensioning the springs inside. Given the power of the springs, any work on them was a dangerous business for both the engineers and also for anyone in the vicinity, so only those with experience and expertise could handle such a task.

BELOW German troops shelter beneath a Flak gun on the Eastern Front. Given their proximity to the muzzle, the troops would be well advised to move before the gun fires. *(AirSeaLand Photos/Cody Images)*

Recuperator cylinder

The recuperator cylinder required regular inspection for two main points: (a) that the gas pressure in the system was high enough to operate the recuperator mechanism and (b) that there was sufficient liquid within the system. The gas pressure could be read on a gas valve: normal pressure was around 600psi. There was also a mechanical method of checking the gas pressure in the system, by raising the gun to its maximum elevation then knocking the gun out of battery and locking the gun back in the full recoil position. When the gun was released from this position, the recuperator cylinder should have provided the motive force to drive the gun back to battery, all being well with the gas pressure.

Checking the liquid level involved dropping the gun barrel to −1 degree and opening the drain plug very slightly (one turn). Rather like checking the water content in a household radiator, if liquid immediately started oozing out from the drain plug then the fluid levels were fine; if no liquid appeared, then the system needed refilling.

The process of refilling and recharging the recuperator cylinder first required that the system was drained of both gas and liquid, performed by removing the liquid and gas filling plug and the drain plug, plus opening the gas valve fully to bleed out any gas pressure. To refill the liquid, the gun had to be set absolutely level and with the drain plug in place, then the fluid (about 17 litres/4 imperial gallons) was poured in through the filling plug until it overflowed, at which point the filling plug cap could be replaced. Then a nitrogen gas supply was fitted to the liquid and gas filling plug, using the required adaptor. With the gas valve on the recuperator open by about two turns, the recuperator was then charged to the correct pressure, the gunner reading the pressure on the supply bottle pressure gauge. Once the correct pressure was reached, the gas valve was closed and then the gas line reconnected and the gas and liquid filling plug replaced.

Recoil cylinder

The chief sign that the recoil system might need its gas and fluid replenishing was excessive recoil, beyond the limits of 105cm (41.5in) at 0 degrees elevation and 70.5cm (27.7in) at maximum

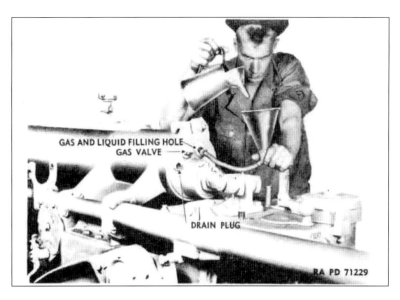

RIGHT **The procedure for filling the recuperator with fluid. Low liquid levels could result in the gun failing to return to battery after the recoil phase.** *(US War Department)*

elevation. The gunners would also look out for signs that the gun was slamming abruptly back into battery, rather than transitioning in a quick but smooth and controlled fashion. Checking that the fluid levels were correct was a simple matter of setting the gun elevation to 2 degrees, then removing the two liquid filler plugs and inspecting to see that the fluid came up to the top of the filling holes. If not, the cylinder's two overflow plugs were also removed, then the system topped up with liquid through the filling holes plus (for final top-up) the overflow holes on the left side of the cylinder, the total system holding about 9.5 litres (2 imperial gallons).

Rammer assembly

The rammer assembly was another of the hydropneumatic systems on the Flak 8.8cm, and testing and maintaining it followed a similar process to that described above for the recuperator. The chief concern for the gun crew was that the gas in the system was at the right pressure (225psi) and that the liquid levels were correct. Checking the former involved connecting a gas pressure gauge to the assembly's gas and liquid filler plug, while checking the liquid involved loosening that plug to see whether liquid oozed out under pressure. The procedure for adding gas and fluid was essentially the same as that for the recuperator. The crew would also need to conduct regular inspections of the rammer tray to ensure that this system was not damaged or defective in any way; the continual impacts of heavy shells on its components was punishing.

Mount and other components

In addition to the major components of the gun itself, the gun mount had to be inspected for the integrity of the fitting between gun and cradle,

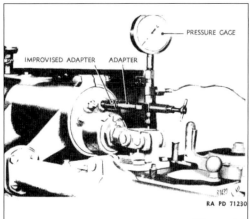

LEFT **Charging the recuperator cylinder with nitrogen gas. Note here that the charger adapter is a US model, this illustration coming from an American technical manual.** *(US War Department)*

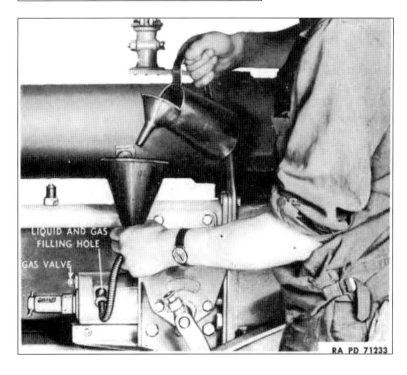

RIGHT **Filling the rammer cylinder with fluid. Liquid was added to the point of overflow.** *(US War Department)*

ABOVE The passage of time shows its evidence in this view of one of the bogie wheels and, at the top of the image, the leaf suspension system.

ABOVE RIGHT Front bogie brake springs; the brake system was placed under intense pressure when the gun was fired from the wheels. *(Author/Axis Track Services)*

with special attention paid to lubricating the pintle bearing and the trunnion bearings. Outrigger hinges had to be properly greased and regularly cleared of rust and dirt build-up. The numerous parts that made up the bogies also needed inspection to ensure that they worked smoothly and had the right levels of lubrication, and that all nuts, bolts and springs were adjusted to the correct tightness. Special attention was naturally paid to the tyres, wheels and brakes. The tyres had to be at the appropriate pressure, unless of course they were solid rubber versions, and the wheels required lubrication with grease, the gunner taking care not to pack too much around the bearings, an excess of which could subsequently leak into the brake drums.

In terms of the brakes, the power brake air line connections had to be examined regularly, as in the harsh conditions of the battlefield it was easy for these cables to become damaged and sustain air leaks, with a subsequent drop in brake pressure. The gunner had to check that the connections between vehicle and bogie brakes were tight and secure, and he would also inspect the connections visually (for cracks) and aurally (for hissing noises). Another useful inspection technique was to cover the cable in soapy water; air leaks would typically produce a soapy bubble at the point of leakage. The handbrake on the rear bogie might also need periodic adjustment by the gun team or by the unit engineers.

RIGHT The rear end of the bottom carriage featured the data transmission junction box. The handle seen to the right is for the levelling jack. *(US War Department)*

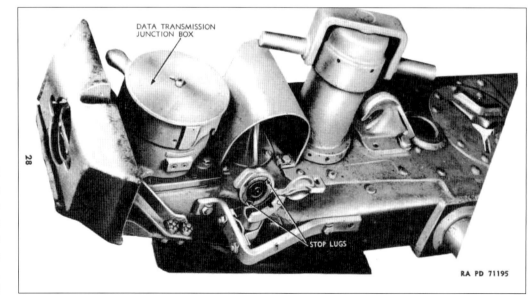

GENERAL GUIDELINES FOR CLEANING AND MAINTENANCE FROM *8.8CM FLAK DESCRIPTION AND OPERATION* (1940)

The gun is to be protected as much as possible from dust, dirt and moisture. Therefore when marching, the muzzle cap, the breech cover and the cover for the fuse setter are always to be in place.

After every use the gun is to be cleaned carefully. As often as the cleaning requires it, but at least after every shot, the breech is to be dismantled to the extent allowed by L.Dv.436, p. 21. Before every firing the breech parts are to be freed as much as possible from dust and windblown sand again. When the breech parts are lubricated, so sulfur solution is to be added to the spindle oil (green) to decrease any occurring erosion.

Dust, dirt and old lubricant are to be removed from all parts, especially on the fuse setter and all sliding surfaces. Barrels are to be cleaned, as a matter of principle, 'immediately' after every firing and then two or three days in a row to avoid build-up of precipitates. The use of strong cleansing agents is forbidden; only prescribed oils and greases may be used.

Heavy copper deposits and hardened substances remaining in corners and on surfaces may be removed only by weapon-technical personnel.

After every cleaning, all exposed parts are to be protected from rust by a thin film of oil or grease. The barrel hooks and sliding rails are to be oiled at short intervals, but at least once after every firing. The open ends of the cradle rails are to be kept clean and constantly to be covered with a film of oil.[45]

Diagnostics

The Flak 8.8cm guns were rugged pieces of kit all round. Between them they put millions of shells into the air or across the landscape of all fronts without excessive rates of failure or problems compared with any other artillery piece. The one major troublesome area of the weapon was the four-section RA 9 bore liner, which, as noted in Chapter 1, contributed towards shell extraction difficulties. Part of this problem was also due to the quality of war-production ammunition; indeed the failures or stoppages that did occur were often due to faulty ammunition as much as a problem with the gun. Other malfunctions were likely due to

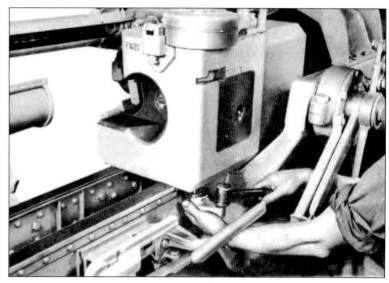

BELOW Removing the breech operating crank, the first stages in the process of breech disassembly. *(US War Department)*

RIGHT Removing the extractor from the breech-block during breech disassembly. *(US War Department)*

the intrusion of dirt, sand or other debris, or from direct battle damage to the parts of the gun.

Whatever the source of the problem, the gun teams had to be conversant in quickly ascertaining the cause of the trouble, and clearing it if possible.

Two faults that posed a particular danger to the gun crew were the failure of the breech-block to close fully, and the apparent non-ignition of a shell on firing. In the case of the former, the two likeliest causes were dirt in the breech preventing the sliding block from moving fully to the closed position, or that during the loading procedure the warhead of a shell was pulled out slightly from the case, thereby increasing the overall length of the shell, with the base of the case preventing breech closure. The manual instructions for these issues flag up the 'Caution!' warning in a bold command.

With the gun in this position, if the breech was closed forcefully by hand, the gun could fire before the operator had time to take his arm away – recoil-smashed hands, arms and shoulders were one of the perennial perils of the artilleryman. Instead it was recommended that the soldier attempt to tap in the breech-block by hammering against it using a block of wood, the wood providing some safety in distance.

In the case of a shell not firing, the gunners had to follow several lines of enquiry to get to the root cause. For example, the problem could actually lie with the firing mechanism spring or the firing pin, which might be weakened, broken or deformed – in which case the parts would need replacing. More dangerously, if proper impact on the primer was attained but the shell still refused to fire after three attempts, then the problem was likely to be a defective shell primer.

BELOW Destruction on the Eastern Front. Wreckage lies strewn around an 8.8cm Flak battery in Crimea, some shells still in their wooden packing crates. *(AirSeaLand Photos/Cody Images)*

LEFT British troops advancing through western Europe pass an abandoned 8.8cm Flak. Note the indirect-fire sight atop the recuperator. *(AirSeaLand Photos/Cody Images)*

BELOW The key firing controls for the Flak 18/36/37. During the counter-recoil phase, the breech was opened, the percussion mechanism cocked and the cartridge case extracted. *(US War Department)*

Again, caution was required here. L.Dv.486 advises in this case:

When the shot does not fire with automatic firing, first determine by looking, not by feeling, if:

1. The right butting face of the wedge is even with the base or protruding a little, and the grip in the spring housing is not fully seated in; or if

2. The right abutting face of the wedge clearly protrudes several millimeters above the base.

In case #1, special care is to be taken, as the shot can accidentally be fired by a careless bump against the thrust crank or the grip of the spring housing. The breech is to be fully closed by a slight bump with a wooden stick (e.g. the handle of a sledgehammer) angled from the front against the grip of the spring housing, whereon the shot will fire.

In case #2, special care is likewise used as the breech will often fully close by a light bump against the wedge and thereby unexpectedly fire the shot. In no case pivot the loading tray inwards or close the breech by hand; rather at first attempt to close the breech by hand with a hit from a wooden stick against the right abutting face

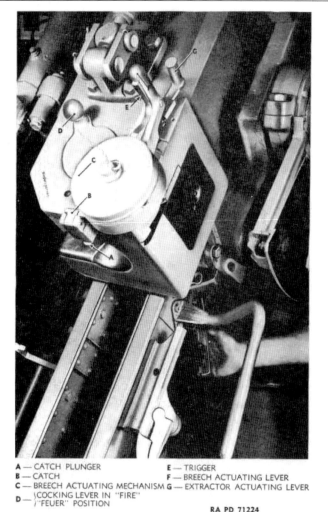

A — CATCH PLUNGER
B — CATCH
C — BREECH ACTUATING MECHANISM
D — COCKING LEVER IN "FIRE"/"FEUER" POSITION
E — TRIGGER
F — BREECH ACTUATING LEVER
G — EXTRACTOR ACTUATING LEVER

RA PD 71224

TRANSPORTATION AND MAINTENANCE

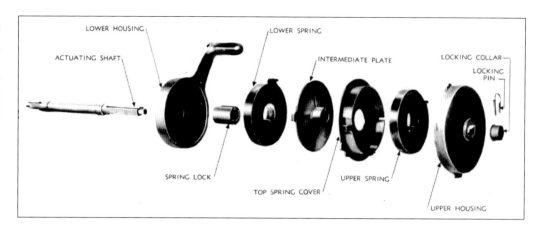

RIGHT An exploded view of the breech actuating mechanism, which is responsible for the manual opening and closing of the breech. *(US War Department)*

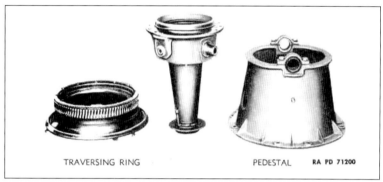

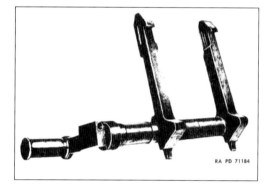

ABOVE The three major components of the 8.8cm Flak gun pedestal. The piece in the centre is the levelling universal, responsible for levelling the emplaced gun precisely. *(US War Department)*

ABOVE RIGHT The shell extractors of the 8.8cm Flak, shown connected to their actuating shaft and operating handle. *(US War Department)*

BELOW A front view of the equilibrators, clearly showing the telescopic housing in which the equilibrator springs moved. *(Author/Muckleburgh Collection)*

of the wedge. If this does not succeed, the breech is very dirty or the projectile with inserting has been a bit loosened in shell, so that the cartridge case protrudes too far rearward. In this case, wait 1 minute, crank the barrel to 15 degrees [angle] and carefully unload.[46]

The advice given here on 'Dealing with duds' assumes that the problem lies with the seating of the shell or the proper closure of the breech, rather than with the shell itself. Alternatively, what could actually be occurring is a 'hangfire', a nightmarish situation in which the primer has been ignited but, because of a defect, is actually slowly smouldering rather than detonating fully and instantly. The problem with a hangfire is that the primer might then suddenly trigger shell-firing, seconds or even minutes after the initial primer strike. This is partly acknowledged in the text above when it recommends waiting one minute before unloading the gun. The US Army manual on the 8.8cm Flak was even more cautious – 'The breechblock will not be opened until at least 10 minutes after the last unsuccessful attempt to fire the piece. The gun will be kept directed in elevation and traverse either on the target or on a safe place in the field of fire.'[47]

In the case of a shell that had fired, but where the spent case remained stuck in the chamber, the failure could be either a ruptured or otherwise deformed case, that the extractor system had broken or that there were some deformities or damage to the breech that prevented the case from smoothly running out. Either way, the spent case had to be knocked out of the open breech using a rammer from the muzzle end, then all parts inspected to ascertain the cause of the problem.

The difficulties outlined here are those related to the gun and its firing mechanism, but there was a whole host of other issues related to the mount, and particularly the recuperator and recoil mechanism. For example, if the gun failed to return to battery after recoiling, the problem could be that there was insufficient gas and liquid pressure in the recuperator, and the levels/pressure needed topping up again. This issue with the recuperator could also explain several other issues, including the gun recoiling more than its maximum allowed distance at specific elevations – there was an *Achtung* mark on the recoil gauge – and also that the gun slid out of battery when it was in an elevated position. Yet the failure to return to battery could be accounted for by several other problems, such as damaged recoil slides or piston rod, or that the recoil mechanism was defective. As always, the gunners were required to be mechanical detectives when the 8.8cm Flak failed to perform as intended.

The Flak 8.8cm would not have entered the list of the Allies' most feared weapons if it had been mechanically unreliable. As it was, properly maintained the Flak gun could keep pumping out shells hour after hour. A masterpiece of engineering, the Flak was nevertheless, in essence, a killing machine. While this book has indulged the understandable fascination with the 8.8cm Flak's engineering, at the same time we must remember that tens of thousands of human beings did not see life beyond the war years because of the capabilities of this formidable weapon.

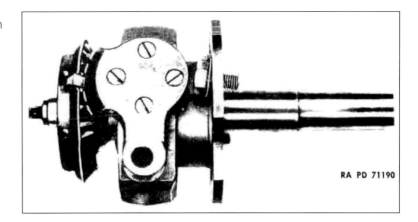

ABOVE The recoil control linkage seen here was the component responsible for altering the length of recoil when the gun was at different elevations. *(US War Department)*

LEFT In maintenance, any toothed components (such as this elevation rack) needed to be kept clean of dirt, debris and ice. *(Author/Axis Track Services)*

LEFT A Flak unit in front of the Acropolis in Athens, May 1941. *(AirSeaLand Photos/ Cody Images)*

Endnotes

1. Nordyke 2005: 685–6.
2. Hogg 1997: 170–1.
3. This chapter quotes extensively from the War Department manual, as it constitutes one of the most useful, accurate and accessible English-language sources for the Flak 8.8cm guns.
4. US War Department 1943: 57.
5. US Army 1943: 19.
6. US Army 1943: 21.
7. US War Department 1943: 32.
8. US War Department 1943: 38.
9. US Army 1943: 138–9.
10. US War Department 1943: 58–59.
11. US War Department 1943: 142.
12. US War Department 1943: 113.
13. US Office of Chief of Ordnance 1945: 185.
14. US War Department 1943: 133.
15. Hogg 1997: 260.
16. Hogg 1997: 142.
17. US War Department 1943: 107–8.
18. Quoted in Anderson 2015: 36.
19. Piekalkiewicz 1992: 123.
20. Boog, Krebs and Vogel 2015: 225
21. US War Department 1943: 43–55.
22. Neumann 1941: 34.
23. Toppe 1991: 89.
24. Toppe 1991: 63.
25. US War Department 1943b:156–57.
26. US War Department 1943b: 156.
27. Neumann 1941: 15.
28. US War Department 1943c. Available at: http://www.lonesentry.com/articles/ttt/german-antiaircraft-defense-flak.html.
29. Krueger 2000: 29.
30. Hammel 2007: 245.
31. Data from Gander: 2012.
32. Here I am reliant upon the detailed analysis in Westermann: 2001.
33. Westermann 2001: 286–87.
34. Westermann 2001: 279.
35. US War Department 1942.
36. US War Department 1942.
37. Quoted in Piekalkiewicz 1992: 42–43.
38. Piekalkiewicz 1992: 60.
39. Piekalkiewicz 1992: 103.
40. US War Department 1943d: 'Tactical Employment of Flak in the Field', *Intelligence Bulletin*, November 1943. Available at: http://lonesentry.com/articles/flak/index.html
41. US War Department 1944. Available at: http://www.lonesentry.com/articles/ttt/german-antitank-tactics.html.
42. Canadian Army Commanders 2005: 6–7.
43. Canadian Army Commanders 2005: 69.
44. US 37th Tank Battalion 1944.
45. Quoted in Piekalkiewicz 1992: 181.
46. Luftwaffe 1936–38: 17–18.
47. US War Department 1943: 77.

Appendix 1

Bibliography and further reading

Anderson, Thomas (2015). *Ferdinand and Elefant Tank Destroyer*. London: Bloomsbury.

Boog, Horst, Gerhard Krebs and Detlef Vogel (2015). *Germany and the Second World War: Volume VII: The Strategic Air War in Europe and the War in the West and East Asia, 1943–1944/5*. Oxford: Oxford University Press.

Canadian Army Commanders (2005). 'The Technique of the Assault: The Canadian Army on D-Day: After-Action Reports by Commanders', *Canadian Military History*, Vol. 14, Issue 3, Article 5.

Ellis, Chris with Peter Chamberlain (1998). *The 88: The Flak/Pak 8.8cm*. London: Parkgate Books.

Gander, Terry (2012). *The German 88: The Most Famous Gun of World War II*. Barnsley: Pen & Sword.

Hammel, Eric (2007). *Aces Against Germany: The American Aces Speak*. Pacifica, CA: Pacifica Military History.

Hogg, Ian V. (1997). *German Artillery of World War Two*. London: Greenhill Books.

Krueger, Lloyd (2000). *Come Fly with Me: Experiences of an Airman in World War II*. Bloomington, IN: iUniverse.

Luftwaffe (1936–38). *8.8cm Flak 18: Description, Method of Operation, and Handling*. Trans. by John Baum. Lisbon, OH: Open Word Berlin 35.

Neumann, Ernst (1941). *8.8 and 2 cm Flak Crew Handbook. Handbook for the Field Artillery Crew (The Cannonier): Weapon and Instruction of the Flak Battery*. Trans. by John Baum. Lisbon, OH: Open Word Berlin 35.

Nordyke, Phil (2005). *All American, All the Way: The Combat History of the 82nd Airborne Division in World War II*. Minneapolis, MN: Zenith Press.

Norris, John (2002). *88 mm Flak 8/36/37/41 & Pak 43 1936–45*. Oxford: Osprey Publishing.

Piekalkiewicz, Janusz (1992). *The German 88 Gun in Combat: The Scourge of Allied Armor*. West Chester, PA: Schiffer Publishing Ltd.

Price, Alfred (1997). *The Luftwaffe Data Book*. London: Greenhill Books.

Toppe, Major-General Alfred (1991). *Desert Warfare: German Experiences in World War II*. Fort Leavenworth, KS: US Army Command and General Staff College.

US Army/Air Force (1953). *German Explosive Ordnance (Projectiles and Projectile Fuzes)*. Washington DC: US Government Printing Office.

US Office of Chief of Ordnance (1945). *Catalog of Enemy Ordnance*. Washington DC: US Office of Chief of Ordnance.

US 37th Tank Battalion, 4th Armored Division (1944). Extracts from after-action report. Available at: https://abramsstandards.com/wp-content/uploads/2016/04/4-WW-II-Excerpts-from-After-Action-Report-37th-Tank-Battalion-4th-Armored-Division-1944.pdf.

US War Department (1942). 'A Tactical Study of the Effectiveness of the German 88 mm Anti-Aircraft Gun as an Anti-Tank Weapon in the Libyan Battle', *Tactical and Technical Trends*, No. 1, 18 June 1942.

US War Department (1943). TM E9-369A, *German 88-mm Antiaircraft Gun Material*. Washington DC: US War Department.

US War Department (1943b). *German Winter Warfare*. Washington DC: Military Intelligence Division.

US War Department (1943c). 'The Organization of German Antiaircraft Defense', *Tactical and Technical Trends*, No. 28, 1 July 1943.

US War Department (1944). 'German Antitank Tactics: Text of a Captured Document', *Tactical and Technical Trends*, No. 51, October 1944.

Westermann, Edward B. (2001). *Flak: German Anti-Aircraft Defenses, 1941–1945*. Lawrence, KS: University Press of Kansas.

Appendix 2

8.8cm Flak 36 gun specifications

Select specifications (with metric conversions) from TM E9-369A, *German 88-mm Antiaircraft Gun Material*.

Gun	
Type	Tube and loose 3-section liner
Total weight	1,337kg (2,947lb)
Weight of removable components:	
Breech ring	229kg (506lb 8oz)
Outer tube	356kg (785lb)
Inner tube	365kg (805lb 8oz)
Liner (muzzle section)	272kg (600lb)
Liner (centre section)	90kg (199lb)
Liner (breech section)	26kg (58lb)
Retaining rings	15kg (34lb)
Overall length of tube	470cm (185in)
Overall length of gun and tube	493.8cm (194in)
Length in calibres	56
Distance from centre line of trunnions to breech face	16.5cm (6.5in)
Travel of projectile in bore	400cm (157.4in)
Volume of chamber	3,703cu cm (226cu in)
Rated maximum powder pressure	33,000psi approx.
Muzzle velocity	*c*850m/sec (2,788ft/sec)
Maximum range:	
Horizontal	14,813m (16,200yd)
Vertical	11,887m (39,000ft)
Maximum effective ceiling	7,620m (25,000ft) at 70 degrees elevation
Rifling:	
Length	400cm (157.4in)
Direction	Right-hand
Twist	Increasing 1 turn in 45 calibres to 1 turn in 30 calibres
Number of grooves	32
Depth of grooves	1mm (0.0394in)
Width of grooves	5mm (0.1969in)
Width of lands	3mm (0.1181in)
Type of breech mechanism	Semi-automatic horizontal sliding block
Rate of fire	15 rounds per minute (practical rate at a mechanised target)
	20 rounds per minute (practical rate at an aerial target)

Recoil mechanism	
Type	Independent liquid and hydropneumatic
Total weight	237kg (524lb)
Weight of recuperator cylinder	129kg (285lb)
Weight of recoil cylinder	108kg (239lb)
Weight of recoiling parts in recoil mechanism	81.9kg (108.5lb)
Total weight of recoiling parts (with gun and tube)	1,433kg (3,159lb)
Type of recoil	Control rod type with secondary control rod type regulating counter-recoil
Normal recoil:	
0-degree elevation	105cm (41.5in)
25-degree elevation	85cm (33.46in)
Maximum elevation	70cm (27.75in)
Capacity of recoil cylinder	9.4 litres (2.5gal)
Capacity of recuperator cylinder	17 litres (4.5gal)
Mount	
Weight (less cannon and recoil mechanism)	3,811kg (8,404lb)
Maximum elevation	85 degrees
Maximum depression minus	3 degrees
Traverse	360 degrees
Loading angles	All angles
Height of trunnion above ground (firing position)	1.58m (5.2ft)
Height of working platform (firing)	0.24m (0.8ft)
Height of trunnion above working platform	1.34m (4.4ft)
Overall dimensions in firing position:	
Length	5.79m (19ft)
Height	2.1m (6.9ft)
Width	5.14m (16.87ft) w/outriggers
Overall dimensions in travelling position:	
Length	7.77m (25.5ft) w/drawbar
Height	2.4m (7.9ft)
Width (front)	2.19m (7.2ft)
Width (rear)	2.31m (7.60ft)
Length of outriggers	1.46m (4.8ft)
Number of bogies	2
Type of bogies	Single axle. Single wheels on front; dual wheels on rear
Weight of front bogie	827kg (1,825lb)
Weight of rear bogie	1,119kg (2,645lb)
Wheelbase	4.19m (13.75ft)
Type of brakes	Vacuum air brakes on all wheels; hand-operated parking brakes on rear wheels also
Type and number of jacks	4 jacks integral with mount for levelling bottom carriage; one on each end of outriggers and carriage
Levelling	4.5 degrees levelling either side of horizontal
Road clearance	0.35m (1.14ft)
Tread (front)	1.77m (5.8ft)
Tread (rear)	1.83m (6ft)
Height of axis of bore above ground (firing)	1.52m (5ft)
Time to change from travelling to firing position	2 minutes 30 seconds with 6-man crew (approx.)
Time to change from firing to travelling position	3 minutes 30 seconds with 6-man crew (approx.)
Weight of entire carriage	7,404kg (16,325lb)
Type of equilibrators	Spring type with built-in spring compressors

Index

Aberdeen Proving Ground, USA 13-14
Acoustic (sound) locators 106-107, 116, 120
Aiming (target acquisition) – see also
  Sighting, and Target detection 48, 79, 99, 110-111
  aiming assembly mechanism 50
  aiming circle device 54, 84
  barrel orientation errors 111
  deflection mechanism 50
  director devices 48
  flight time of shells 48, 110
  night-time 22, 48, 61, 114
  optical distortion 93
  range elevation and calculations 118
  range estimation 48, 91-92
Aircraft
  Boeing B-17 Flying Fortress 114, 116; B-17G 18-19
  Douglas C-47 143
  Junkers Ju 87 Stuka 125
  Martin B-26 Marauder 102
  North American P-51 Mustang 116
Allied bombing raids 78, 102, 104, 112, 114, 117, 120
  aircraft losses 119-120
  day bombing 119-120
  night bombing 119-120
  US aircraft damaged 121
Ammunition – see also Shells 22, 57-73, 126
  anti-tank warheads 67-71
  armour piercing (AP) 23, 59, 62-65, 68, 71-72, 125-126, 128, 130-131, 133
  brass-cased 19, 60-62
  burst radius 64, 110
  bursting charge 65
  drill rounds 79
  condition 99
  driving bands 67-68, 72-73, 99
  faulty (duds) 161-162, 164
  fragmentation 64
  grooved (*Gerillt*) 65
  high-explosive (HE) 13, 57-58, 62, 64-66, 70-71, 110
  high-explosive anti-tank (HEAT) shell 68
  hollow charge 68, 70
  incendiary 65
  percussion-type 59
  ranging shells 94
  shell cases 58, 127, 165
  shell identification marks 60, 66
  steel-cased 19, 60
  tropical shells 61
  warheads 64-71
Ammunition cradles and storage lockers 27, 69, 111

Anti-aircraft (AA) guns and operations – see also Flak 8.8cm guns 7, 10-11, 18, 22, 79, 91, 101-121
  Ehrhardt BAK 11
  Flak guns (2cm/3.7cm/5cm/10.5cm/12.8cm) 102-104, 109, 115, 140, 152
  Gerät 37 19
  Gerät 42 21-22
  Krupp L/35 10
  Krupp L/45 11
  Krupp/Bofors L/60 12
  machine guns and cannon 102
  night-time work 103, 105
  Pak 36 and 38 111
  Rheinmetall Kw-Flak L/27 10
  Rheinmetall L/45 11
Anti-balloon cannon 10
Anti-naval traffic 22, 27, 135, 142
Anti-tank (AT) operations 6-7, 16, 18, 21-24, 87, 91-92, 103-104, 117, 122-147
  tactics 142
  tank kills 133, 135, 139
Auchinleck, Gen Claude 129
Anzio bridgehead 143
Armoured train 27

Barrels 18-19, 29-32, 98
  bore inspection 156
  explosions 105, 140, 156
  inner liners 15, 18, 30-31, 156, 161
  installation 92
  kill rings 122, 139
  lifespan 61, 72
  locking collar and arm 30, 42, 61, 135
  multi-section 19
  muzzle 144, 158
    brake 24
    covers and plugs 91, 94, 98, 145
    rest and lock 32, 77, 135
  one-piece (monobloc) 10, 14-15, 18, 24, 30
  RA 9 multi-section 14-15, 18-19, 30, 61, 161
  rifling 23, 30, 67, 70, 94
  service life 14-15
  stepped 14
Blitzkrieg 124
Bofors 11-12
Borgward 150
Bravery awards 134
Breech 16, 23-24, 30-32, 45, 86, 147, 156-157, 161, 163
  block 24, 30, 32, 34, 36, 60, 94, 97, 161-162, 164
  cocking lever 36, 124, 156
  cover 137

loading 45, 58, 79
loader mechanisms 18, 20
loading tray 16, 35, 45, 69
mechanism 34, 64, 94, 97, 99, 156-157, 164
operating handle 35
rammer system 45, 157, 159, 165
recuperator 29, 36-37, 98, 158-159, 163, 165
shell extractor mechanism 35, 161, 164-165
trigger mechanism 42
British Army 141
  4th Armoured Brigade 128
  22nd Brigade 129
  11th Hussars 130

Calibre 10-11
Camouflage 77, 80, 90-91, 113, 121, 127, 140-141
Captured guns 10, 12, 23, 122, 127
Cavalry Tank Museum, India 18
Chaff 116
Churchill, Winston 130
Cleaning and maintenance 94-99, 154-165
  daily 154-155, 157
  dust and dirt protection 91-92, 137, 162
  General Guidelines 161
  logbooks 155
  lubricating 94, 99, 156, 158, 160
Communications 13
  data transmission receptacle 47
  gun receiver systems 48
  radio headsets 17, 55, 87
  telephone systems 53, 86-87
  wireless 125
Condor Legion 6, 16
Crews (Flak teams) 48, 75-99, 113, 125, 128, 130, 132, 134-135, 139-140, 143, 147, 149, 154-155
  armour protection 153
  casualties 135
  director team 112
  disease and heatstroke 91
  drinking 91
  fire direction team 114
  firing tempo 94
  flak helpers 78-79
  gunlayers 48, 85
  gun leaders 80, 85-86, 88-91, 154
  gun loaders 109, 141, 147, 155
  human factors 99
  inexperienced 135
  Kanoniere (cannoniers) 80, 96
  key qualities 96

oath of service 109
personalised weapons 76
rearranging priorities 87
skeleton 91
Slovak insurgents 79
training and manuals 78-80, 84, 87-88, 96, 98, 131
transferred to field service 121

Daimler-Benz 150, 152
Data transmission junction box 160
Detonation altitude 16, 110-111
Diagnostics 161-165
Displayed guns 6, 12, 18, 19, 22

Ehrhardt, Heinrich 10
Electrics and lights 46, 53, 88, 129
Elevation – see Traverse and elevation
Emplacements and setting up 80, 82-83, 91-92, 97, 126-127, 130
  on frozen ground 97-98
Explosives abbreviations 66
Exports 12

Fire control systems 13, 20, 22, 30, 47, 49, 51-54, 84, 86, 109-118, 163
  aerial 16
  barrage method 114
  centralised 11
  continuously pointed fire 112
  failures 162
  fire patterns 112
  intervals 51
  Kdo. Gr. fire directors 52, 55
  Pakfront tactic 135
  portable 54-55
  predicted concentration 113
  radio instructions 17
  receiver system 16-17, 21
  regional 90
Firing 34-35, 45, 88, 94
  cocking and recocking 45, 94, 157
  electric primer 59
  firing mechanism 157, 162, 165
  firing pin 59-60, 65, 94, 97, 157, 162
  'hangfire' 164
  percussion primer 35, 59, 163
  predictor 14, 108-110
  unload (shell extraction) 21, 45, 94, 161
First World War 6, 9-10, 12
Flak batteries 58, 76, 101, 108-110, 112, 117, 121, 124, 131, 162
  Grossbatterien 109
Flak name 10
Flak shrapnel 115
Flak towers 115
Flak unit organisation 103-109
Flak 8.8cm guns
Flak 16 10, 12
Flak 18 L/56 6, 10-18, 22-23, 25-27, 30, 43, 70-71, 88, 95, 109-111, 113, 117, 119, 121, 124, 126-127, 129-131, 133, 135, 137, 140, 145-147, 152-153, 163
Flak 36 6, 12-18, 22, 27, 30-31, 43, 46, 70-71, 76, 147, 152, 163
Flak 36/37 13-15, 116

Flak 37 6-7, 12-18, 21-22, 25-27, 29-30, 39, 45, 69, 71, 79, 89, 97, 102, 147, 152, 163
Flak 37/41 20-21, 70
Flak 41 6, 18-21, 26, 67, 70-71, 147, 152
Franco-Prussian War 1870-1 10
Fuzes 16, 62-64, 126
  base-detonating 63, 68, 72
  handling in action 84-90
  inertia-type 62
  Junghans centrifugal drive 62
  Krupp-Thiel clockwork timer 62, 64
  percussion 62-63, 65, 73
  proximity 63
  setting 79, 86, 88, 96, 111, 121
  timed 62-63, 65, 72, 87-88, 110
  warhead 62, 96
Fuze-setting mechanism 16, 45-47, 86
  receiver 54, 86

German Army (Heer) 14, 78, 104-106, 124-125, 130, 132-133, 136, 138, 144
  Afrika Korps 128
  Weapons Agency 20
  79th Mountain Artillery Regiment 136
  15th Panzer Division 132
  21st[t] Panzer Division 132
German Navy (Kriegsmarine) 10, 104
German War Ministry 11
Ground fighting – see also Anti-tank guns, and Sighting 121, 124-147
Gun bell 18, 51
Guns and artillery pieces – see also Anti-Aircraft guns, and Flak 8.8cm guns
  Bazooka 68
  Big Bertha 6
  Kanone 18 15cm 152
  KwK 36 24
  KwK 43 tank gun 22, 72-73, 140, 146
  L/45 naval gun 10
  M1 90mm 147
  Morser 18 21cm 152
  Oerlikon cannon 141
  Pal 35/36 125
  Pak 41 23
  Pak 43 anti-tank 22-24, 71-73, 140, 144; Pak 43/1 25; Pak 43/2 L/71 7, 25-26; Pak 43/3 L/71 26
  Pak 43/41 22-23
  Pak 44 L/55 26
  PIAT anti-tank 24
  tank guns 22, 71-73
  tank destroyer guns 72, 143
  3.7in Mk 3 147
  105mm assault gun 146
  17pdr 147
Gun shields 19-20, 23, 25, 46, 125, 133, 153

Hammel, Eric 116
Hass, FM Karl 133
Hitler, Adolf 135, 137
Hogg, Ian V. 19-20, 60, 64
Hussein, Saddam 6
  Supergun 6

Jones, Pte Gerald D. 6-7

Kesselring, General-FM 143
Kirst, Kanonier Wolfgang 76
Koch und Kienzle 11-12
Krauss-Maffei AG 150
Krueger, US airman Lloyd 114
Krupp 10-12, 16, 22

Lewis, Pte First Class 6
Luftwaffe 18-19, 21, 104-105, 109, 126, 134, 136, 144
Luftwaffe Flak branch 103, 120, 124-125
  3rd Flak Division 132
  Flak Regiment 22 124
  Flak Regiment 135 132
  soldiers 77, 80

Maintenance – see Cleaning and maintenance
Maybach petrol engines 151-152; V-12 26
Mediterranean Allied Air Forces (MAAF) 120
  Ninth Air Force 120
  Twelfth Air Force 120
  Fifteenth Air Force 120
Monte Cassino 147
Mounts 30, 38-47, 98, 126, 159-160, 169
  carriage 6. 16, 29, 98, 127; lower (bottom) 39-41, 151, 160; rear 69; upper, 38, 41-47, 62
  carriage-levelling mechanism 41, 43, 83
  coastal craft 27
  cruciform platform 13, 15, 23, 43, 126
  equilibrator 29, 37, 164
  ferry platforms (Siebel Ferry) 27
  gun cradle 42
  gun winch mechanism 76
  outriggers and levelling jacks 15, 38-41, 43, 48, 76, 81-83, 93, 97-98, 110, 144, 160
  turntable 19
Muckleburgh Military Collection 7, 13, 124
Mueller, Lt Conrad 145
Museo Histórico Militar de Cartagena, Spain 18

Neumann, Ernst 84, 87-88

Operation Barbarossa 23, 133, 136
Operation Crusader 129
Operation Overlord 143

Paris 145
Performance 119
  ceiling/altitude 13, 21, 62
  effectiveness 119
  kill rates 112
  muzzle velocity 10-12, 18, 20-21, 48, 59-60, 70, 72
  Pak 43 23
  range 55, 62, 139
  rate of fire/rounds per minute 14, 18
  total aircraft kills 119-120
Personnel engaged in AA work 78, 103
PoWs 78-79
Ploesti oil refineries 120
Porsche, Ferdinand 25
Production figures 21, 23, 26, 119
  halftracks 25, 150, 152, 154

Propaganda 14, 112
Propellants 14, 58-61, 71, 89
  double-bass 61

Radar 48, 109, 111, 114
Rangefinders 16, 44, 48, 52, 55, 77, 92, 111
  coincidence 134
  optical 11, 54
  stereoscopic 53, 55, 117, 120
Recoil 36-38, 94, 97, 124, 141
Recoil mechanism 29-30, 35-38, 98-99, 158-159, 165, 169
  brake 99
  counterrecoil mechanism 99
  cylinder 29, 37-38, 158
  indicator 124
  liquid 98, 158
  marker scale 37, 98
Red Army 136, 156
Reich Labour Service 78
Rheinmetall 10-11, 14, 16, 18
Rheinmetall-Borsig 10
Rommel Gen. Maj. Erwin 125-126, 132
Royal Air Force (RAF) Bomber Command 105, 119
Royal Canadian Artillery 12
  14th Field Regiment 143

Saurer Werke 150
Searchlights 22, 79, 106-107, 114, 116
Self-propelled (SP) guns 23-26, 73
  SdKfz *Hornisse/Nashorn* 24-25
SdKfz 8 25-26, 131, 152
  VFW Pz Sfl IVC 25
Shell types – see also Ammunition
  Gr Patr 39 HL 70
  H1 Gr Patr 39 Flak L/4.7 70
  KwK 43 L/71 59, 72
  Leuchtgeschoss L./4.4 70
  Pzgr Patr 39 68-69, 71, 130
  Pzgr Patr 40 69, 72
  Pzgr 39/1 72
  Pzgr 39/43 68, 72
  Sprgr Flak 41 67, 70
  Sprgr Patr L/4.4 HE 58
  Sprgr Patr L/4.5 63-65, 71, 73; internal component 61
  Sprgr Patr L/4.7 FES 67, 73
Sighting – see also Aiming 24, 30, 47-51, 88
  aerial targets 49, 94
  battery commander's telescope 54-55
  direct-fire shooting 23, 49-51
  elevation quadrant 44, 47, 50, 89
  ground targets 50-51, 84, 88, 94
  indirect fire sight 163
  optical 49, 91, 137-138
  panoramic telescopes 31, 52
Stereoscopic Director (*Kommandogerät*) 36 112, 114, 117-118, 120
  wind correction knobs 118
  telescopic sights 14, 21, 47, 49, 51, 88
  tracking telescopes 52
Spall 67
Spanish Civil War 6, 13-14, 16, 84, 124

Specifications 21
  Flak 36 gun 168-169
  SdKfz 7 152
Stalingrad 136-138
Swedish Army 12

Tanks and tank destroyers 26
  Elefant 7, 26
  German Marks III and IV 128, 144
  Jagdpanther 26
  KV-1 2
  KV-2 134
  Matilda I and II 125-126
  M10 6-7
  Sherman 142, 144
  SOMUA S35 125
  Panther 26, 144
  Tiger I 23-24, 73, 144
  Tiger II 24, 140, 146
  T-34 23, 48, 133-134, 136
Target detection 106-107, 117, 134
  range bearing 106, 118
  target course plate 118
Theatre challenges 90-99
  desert operations 91-95, 97
  dust 94-95
  sandstorms 95
  winter operations 95-99, 135
Theatres of operation 103, 130
  Atlantic Wall 22
  Balkan 27, 95
  Baltic 95, 135
  Belgium 59, 125
  Berlin defence 21, 78, 112
  Crimea 136, 162
  Czechoslovakia 114
  D-Day landings, Normandy 95, 141, 143-144
    Omaha Beach 143
    Utah Beach 144
  Eastern Europe 95
  Eastern Front (Russia/Soviet Union) 22, 25, 63, 76-78, 91, 95-97, 103, 132-138, 141, 147, 156, 158, 162
  East Prussia 95
  Egypt 127
  European Theatre of Operations (ETO) 140
  France 18, 25, 58, 80-81, 103, 105, 113, 116-117, 125-126, 140, 144, 146
  Greece 114, 117, 165
  home defence, Germany 20, 78-79, 103-105, 107-108, 112, 115, 119, 120-121, 147
  Hungary 25, 85
  Italy 18, 24-25, 90, 95-96, 120, 138-140, 143, 147
  Libya 93, 127
  Mediterranean 27
  Netherlands 140
  North Africa 18, 20, 61, 84, 91-92, 124, 126-133, 138-139
  Northern Europe 95, 145
  Poland 119, 124-125
  Sicily 138
  Tunisia 60, 130, 146

Ukranian steppe 95
Western Desert 18, 95, 126-132
Western Europe 124-125, 163
Western Front 11, 113, 141-144, 147
Third Reich 7, 108, 119, 121, 135, 137, 150
Tobruk 128-129
Tools 76, 149, 155
Toppe, Maj-Gen Alfred 91, 93-94
Transportation 6, 15-16, 18, 23, 25, 29-30, 40, 42, 46, 53, 85, 93, 95, 138, 149-154
  bogies and wheels 17, 39-40, 81, 93, 97, 125, 127, 132, 160
  Büssing-NAG BN 9 154
  Daimler-Benz DB10 halftrack 25
  halftracks 25, 93, 149-154
  Hanomag SS-100 154
  horse-drawn wagons 138
  railway wagons (Flak Trains) 27, 101, 150
  SdKfz 7 69, 130, 145, 149-152
    Bunker Cracker 153
    driver's compartment 152
    gunner's seats 21, 80, 89, 151
    towing connection 83, 97, 154
  SdKfz 9 halftrack 25
  Sonderanhänger 201 15-17, 40, 95
  Sonderanhänger 202 14, 39, 48, 54
  Vomag truck 16
Traverse and elevation 16, 43, 50, 52, 55, 86-87, 90, 99, 157-158
  angle of site mechanism 50
  azimuth receivers 47, 53
  elevating mechanism 42-44, 48
  elevation micrometer 138
  elevation receivers 16, 31, 47, 50, 53, 88, 99
  testing 157
  traversing mechanism 16, 18, 21, 42-43, 99, 137

USAAF 119
  8th Air Force 119, 121
  15th Air Force 121
  9th Bombardment Division 102
US Army 37
  82nd Airborne Division 6
  4th Armoured Division 145
  29th Division, 116th Regiment 143
  8th Infantry Division 145
  2nd Rangers 143
  5th Rangers 143
  37th Tank Battalion 145
US Army *Catalog of Enemy Ordnance* 52
US War Department Technical Manual 43, 50, 53, 71, 82, 91, 118, 164

Versailles Treaty 11
V-1 and V-2 rocket sites 104

Warheads – see Ammunition
Waffenfabrik Solothurn AG 11
Waffen-SS 137
Westermann, Edward B. 120-121
Wilde, Capt 6-7
Wolz, Oblt Alwin 132

Zaloga, Steven J. 104